普通高等教育"十四五"计算机类专业系列教材

Jakarta Servlet 开发教程
——基于 Servlet、Spring、SpringBoot 案例

肖 川◎编著

中国铁道出版社有限公司
CHINA RAILWAY PUBLISHING HOUSE CO., LTD.

内 容 简 介

本书针对普通高等教育计算机类专业教学要求编写。本书以 Jakarta Servlet 6.0 规范为导向，以案例为驱动介绍 Web 应用的不同功能模块，用 16 个逐步深化的案例循序渐进地讲解了基于 Servlet 开发 Web 应用的基本概念和设计思路，主要内容包括 Servlet、过滤器、显示静态资源、显示模板网页、非阻塞输出、分派请求、会话、提交表单、文件上传及下载、验证用户输入、国际化、JPA、RESTful 服务、SSE、WebSocket、异常的统一处理。本书对每个案例同时提供基于 Servlet、Spring、SpringBoot 的三种实现方案，以便读者对比三种不同的实现方案并更加深刻地理解不同层次的封装。通过本书的学习，读者不仅可以掌握 Web 开发的底层设计逻辑，还能使用具有更高开发效率的高层封装框架。

本书适合作为高等院校 Java Web 开发相关课程的教材，也可作为各类进修班、培训班讲授 Web 开发的教材，还可作为 Web 应用软件开发者的参考书。

图书在版编目（CIP）数据

Jakarta Servlet 开发教程：基于 Servlet、Spring、SpringBoot 案例/肖川编著.—北京：中国铁道出版社有限公司，2023.10
普通高等教育"十四五"计算机类专业系列教材
ISBN 978-7-113-30480-5

Ⅰ.①J… Ⅱ.①肖… Ⅲ.①JAVA 语言-程序设计-高等学校-教材 Ⅳ.①TP312.8

中国国家版本馆 CIP 数据核字（2023）第 154924 号

书　　名：Jakarta Servlet 开发教程——基于 Servlet、Spring、SpringBoot 案例
作　　者：肖　川

策　　划：翟玉峰　谢世博	编辑部电话：(010)83525088
责任编辑：包　宁	
编辑助理：谢世博	
封面设计：尚明龙	
责任校对：安海燕	
责任印制：樊启鹏	

出版发行：中国铁道出版社有限公司（100054，北京市西城区右安门西街 8 号）
网　　址：http://www.tdpress.com/51eds/
印　　刷：东港股份有限公司
版　　次：2023 年 10 月第 1 版　2023 年 10 月第 1 次印刷
开　　本：787 mm×1 092 mm 1/16　印张：15.75　字数：369 千
书　　号：ISBN 978-7-113-30480-5
定　　价：48.00 元

版权所有　侵权必究

凡购买铁道版图书，如有印制质量问题，请与本社教材图书营销部联系调换。电话：(010) 63550836
打击盗版举报电话：(010) 63549461

前　言

Jakarta Servlet 是一组接口规范，用于支持 Jave Web 应用的开发；诸多 Web 服务器提供了 Jakarta Servlet 的具体实现，这使之具有广泛应用的基础。目前 SpringBoot 框架因其较高的开发效率成为 Java Web 技术选型的首选，这是因为 SpringBoot 框架为了让开发者可以更多地关注业务功能，而对底层 Servlet 技术细节进行了封装，封装层次为 SpringBoot 封装了 Spring，而 Spring 封装了 Jakarta Servlet。封装的好处是界面友好、降低了使用门槛，封装的弊端是缺少对细节的掌控、提高了个性化定制的成本。本书抛开上层的封装框架，直接用底层 Jakarta Servlet 技术进行 Web 应用开发。作为对比，本书亦提供基于 SpringBoot 框架及 Spring 框架的实现方案。

根据教育部新工科建设的方针，高校的课程要对经济发展和产业转型升级发挥引领和支撑作用。在互联网发展到云计算时代的今天，Servlet 规范历经 6 次版本更迭，最新版本发布于 2022 年 5 月 12 日。了解、掌握新的 Servlet 技术并在网站开发中熟练应用使之服务于区域经济建设，这是开设 "Web 网站开发" 课程的目的。目前市面上 Web 网站开发相关的教材或者是介绍 Spring 和 SpringBoot 的，或者是讲解 Servlet 和 JSP 的，鲜见把 Servlet 和 Spring 及 SpringBoot 合在一起的，而事实上这三者的关系密切，因此编写此教材正是填补现有教材未能三者兼顾的空白。

本书特色

本书主要介绍 Web 开发的底层 Servlet 设计逻辑，也兼顾介绍具有更高开发效率的高层封装框架 Spring 和 SpringBoot。在内容选材上，教材注重切合实际需要，选用业界普遍认可的更新的技术模块，比如用 Thymeleaf 代替 JSP、用 JPA 代替 Mybatis，并增加了 SSE 和 WebSocket 等较新的知识，这种选材使得教材的适用范围更广，更具有前瞻性。在课时安排上，本教材适宜每章 2 学时，第 10 章至第 13 章由于内容较多，可以延长至每章 4 学时。本教材的读者应该熟悉 Java 语法、具备 Java 的面向对象知识并对 HTTP 协议有初步的了解。

- 遵循 Jakarta Servlet 6.0 规范。
- 选用 Thymeleaf 作为页面模板引擎，以便于前后端分离。
- 渐进式案例。每一章案例都是在上一章案例的基础上进行扩展，把上一章案例的项目副本作为本章案例项目的起点；每一章案例的实现只描述本章所扩展的内容。
- 每章案例均采用 Servlet、Spring、SpringBoot 三种实现方案。通过对比，可以对各种实现方案有更深刻的理解。

软件版本

本书案例所用软件及 API 的版本如下：
- JDK 17
- Tomcat (embed) 10.1.7
- Spring 6.0.6
- SpringBoot 3.0.4
- MySQL 8.0.31
- Maven 3.6.1
- Jersey 3.1.0
- Jackson 2.14.2
- Jakarta Servlet API 6.0.0
- Jakarta Persistence API 3.1.0
- Jakarta Validation API 3.0.2
- Jakarta rs API 3.1.0
- Jakarta WebSocket API 2.1.0
- Hibernate core 6.1.7.Final
- Hibernate validator 8.0.0.Final
- Thymeleaf 3.1.1.RELEASE

读者对象

- 高校"Java Web 开发"相关课程的学生
- Java Web 开发者
- 其他对 Java Web 开发感兴趣的读者

主要内容

本书分 16 章介绍 Java Web 应用开发技术，每章用一个案例说明其使用场景。每个案例分别用 Jakarta Servlet、Spring、SpringBoot 三种方案实现。由于封装层次不同，案例的不同实现所用方法有所不同。

- 第 1 章

介绍 Servlet 及其相关的基础概念，包括 Servlet 容器、Servlet、ServletRequest、ServletResponse。本章案例是在浏览器窗口显示简单文本。

- 第 2 章

介绍过滤器。本章案例是在过滤器中修改请求参数的值以及设置应答的字符编码。

- 第 3 章

介绍路径映射，把 URL 网址映射到 jar 包内或者文件系统内静态文件。本章案例是在浏览器中通过网址访问 jar 包内或文件系统内静态文件。

- 第 4 章

介绍监听器与 Thymeleaf 模板引擎。本章案例是在浏览器中通过网址访问 HTML 网页。

- 第 5 章

介绍 Servlet 的非阻塞输出。本章案例是以非阻塞的方式输出静态文件及 HTML 网页至浏览器。

❑ 第 6 章

介绍把请求分派给其他 Servlet 处理。本章案例是通过同步分派（forward）或者异步分派（dispatch）把输出 HTML 网页的工作交由专门的 Servlet 完成。

❑ 第 7 章

介绍会话类 HttpSession。本章案例是实现具有会话特性的多次访问。

❑ 第 8 章

介绍数据绑定，包括输入数据绑定和输出数据绑定。输入数据绑定是把表单提交的参数转换成对象属性，输出数据绑定是把对象属性回显至表单元素或者页面元素。本章案例是实现用户信息录入及修改。

❑ 第 9 章

介绍附件上传及下载。本章案例是在录入或者修改用户信息时增加附件上传的功能，并且提供附件下载的功能。

❑ 第 10 章

介绍对象验证，根据所配置的验证注解对某个对象实例进行验证，配置所用验证注解可以是预定义注解也可以是自定义注解。本章案例是在录入或者修改用户信息时增加输入验证功能。

❑ 第 11 章

介绍国际化，包括处理请求时读取地区设置、作出应答时指定地区设置，以及根据请求时的语言参数应答对应语言的页面信息、出错提示信息。本章案例是提供不同语言版本的用户管理功能。

❑ 第 12 章

介绍 Servlet 如何通过 JPA 操作数据库。本章案例是把操作者在浏览器中录入及修改的用户信息保存至 MySQL 数据库，并把从 MySQL 数据库查询到的用户信息在浏览器窗口显示。

❑ 第 13 章

介绍如何用 Servlet 提供 RESTful 服务。本章案例是通过浏览器或者 jersey 客户端访问 Servlet 所提供的 RESTful 服务。

❑ 第 14 章

介绍如何用 Servlet 提供事件推送服务。本章案例是编写一个模拟落球的应用。

❑ 第 15 章

介绍如何用 Servlet 处理 WebSocket 通信。本章案例是编写一个简易聊天室。

❑ 第 16 章

介绍如何对 Servlet 的运行异常进行统一处理。本章案例是用统一的页面显示 Servlet 运行时出现的异常。

源码课件

本书的 Java 源码和 PPT 课件可从中国铁道出版社教育资源数字化平台（http://www.tdpress.com/51eds/）上获取。

致谢

感谢我的家人。本书的写作占用了大量的业余时间，若没有家人的理解和支持，本书不可能完成。

感谢中国铁道出版社有限公司的各位编辑。在他们的鼓励和帮助下，本书才会顺利出版。

由于编者水平有限，书中难免有不足之处，还望读者海涵和指正。非常期待能够得到广大读者的反馈，在技术之路上互勉共进（编著者邮箱：cxiao@fudan.edu.cn）。

肖　川

2023 年 5 月

目 录

第1章 Servlet 1
1.1 相关概念 1
1.1.1 Servlet 容器 1
1.1.2 ServletContext 接口 2
1.1.3 Servlet 接口 4
1.1.4 ServletRequest 接口 5
1.1.5 ServletResponse 接口 ... 6
1.1.6 应答时字符编码 6
1.2 案例描述 7
1.3 用 Servlet 实现 7
1.4 用 Spring 实现 9
1.5 用 SpringBoot 实现 11
1.6 小结 12
1.7 习题 12

第2章 过滤器 13
2.1 相关概念 13
2.1.1 过滤器概念 13
2.1.2 包装器 15
2.1.3 配置过滤器 15
2.2 案例描述 16
2.3 用 Servlet 实现 17
2.4 用 Spring 实现 19
2.5 用 SpringBoot 实现 20
2.6 小结 21
2.7 习题 21

第3章 显示静态资源 23
3.1 相关概念 23
3.1.1 访问 jar 包资源 23
3.1.2 路径匹配 23
3.1.3 ServletResponse 的关闭 ... 25
3.2 案例描述 25
3.3 用 Servlet 实现 26
3.4 用 Spring 实现 27
3.5 用 SpringBoot 实现 28
3.6 小结 29
3.7 习题 29

第4章 显示模板网页 30
4.1 相关概念 30
4.1.1 Thymeleaf 30
4.1.2 ServletContext 命名属性 ... 32
4.1.3 事件与监听器 32
4.1.4 注册监听器 33
4.2 案例描述 34
4.3 用 Servlet 实现 35
4.4 用 Spring 实现 38
4.5 用 SpringBoot 实现 40
4.6 小结 41
4.7 习题 41

第5章 非阻塞输出 42
5.1 相关概念 42
5.1.1 异步输出 42
5.1.2 AsyncContext 43
5.1.3 WriteListener 44
5.2 案例描述 44
5.3 用 Servlet 实现 45
5.4 用 Spring 实现 49

5.5 用 SpringBoot 实现 50
5.6 小结 50
5.7 习题 50

第 6 章 分派请求 51
6.1 相关概念 51
  6.1.1 委托的分类 51
  6.1.2 获取 RequestDispatcher 对象 .. 52
  6.1.3 使用 RequestDispatcher 对象 .. 53
  6.1.4 分派方法的区别 53
6.2 案例描述 55
6.3 用 Servlet 实现 55
6.4 用 Spring 实现 58
6.5 用 SpringBoot 实现 59
6.6 小结 60
6.7 习题 60

第 7 章 会话 61
7.1 相关概念 61
  7.1.1 会话跟踪机制 61
  7.1.2 HttpSession 62
  7.1.3 在 Thymeleaf 的 URL 中
        传递参数 62
7.2 案例描述 62
7.3 用 Servlet 实现 64
7.4 用 Spring 实现 67
7.5 用 SpringBoot 实现 69
7.6 小结 70
7.7 习题 70

第 8 章 提交表单 71
8.1 相关概念 71
  8.1.1 请求时字符编码 71
  8.1.2 输入数据绑定 72
  8.1.3 输出数据绑定 72
  8.1.4 sendRedirect(.)方法 73

8.2 案例描述 73
8.3 用 Servlet 实现 74
8.4 用 Spring 实现 91
8.5 用 SpringBoot 实现 96
8.6 小结 97
8.7 习题 97

第 9 章 文件上传及下载 98
9.1 相关概念 98
  9.1.1 表单的 enctype 属性 98
  9.1.2 Multipart 99
9.2 案例描述 99
9.3 用 Servlet 实现 101
9.4 用 Spring 实现 110
9.5 用 SpringBoot 实现 115
9.6 小结 115
9.7 习题 115

第 10 章 验证用户输入 116
10.1 相关概念 116
  10.1.1 预定义的验证注解 117
  10.1.2 自定义的验证注解 118
10.2 案例描述 118
10.3 用 Servlet 实现 119
10.4 用 Spring 实现 127
10.5 用 SpringBoot 实现 133
10.6 小结 134
10.7 习题 134

第 11 章 国际化 135
11.1 相关概念 135
  11.1.1 请求时地区设置 135
  11.1.2 应答时地区设置 136
11.2 案例描述 136
11.3 用 Servlet 实现 137
11.4 用 Spring 实现 147

11.5 用SpringBoot实现 153	14.3 用Servlet实现 198
11.6 小结 154	14.4 用Spring实现 209
11.7 习题 154	14.5 用SpringBoot实现 211
	14.6 小结 211
	14.7 习题 211

第12章 JPA 155

第15章 WebSocket 213

- 12.1 相关概念 155
  - 12.1.1 JPA 概述 155
  - 12.1.2 基本注解 157
  - 12.1.3 对象的状态转换 158
- 12.2 案例描述 159
- 12.3 用 Servlet 实现 160
- 12.4 用 Spring 实现 168
- 12.5 用 SpringBoot 实现 174
- 12.6 小结 175
- 12.7 习题 175

- 15.1 相关概念 213
  - 15.1.1 WebSocket 概述 213
  - 15.1.2 事件驱动 API 214
- 15.2 案例描述 215
- 15.3 用 Servlet 实现 215
- 15.4 用 Spring 实现 218
- 15.5 用 SpringBoot 实现 222
- 15.6 小结 222
- 15.7 习题 223

第13章 RESTful 服务 176

第16章 异常的统一处理 224

- 13.1 相关概念 176
  - 13.1.1 RESTful 176
  - 13.1.2 非阻塞输入 177
- 13.2 案例描述 177
- 13.3 用 Servlet 实现 178
- 13.4 用 jersey 框架实现 rs-client 187
- 13.5 用 Spring 实现 193
- 13.6 用 SpringBoot 实现 194
- 13.7 小结 195
- 13.8 习题 195

- 16.1 相关概念 224
  - 16.1.1 sendError(.)方法 224
  - 16.1.2 出错处理 Servlet 224
- 16.2 案例描述 225
- 16.3 用 Servlet 实现 226
- 16.4 用 Spring 实现 228
- 16.5 用 SpringBoot 实现 229
- 16.6 小结 231
- 16.7 习题 231

第14章 SSE 196

附录 A 安装及设置 232

- 14.1 相关概念 196
  - 14.1.1 SSE 特点 196
  - 14.1.2 事件队列 197
- 14.2 案例描述 197

附录 B 初始项目 238

附录 C 注解式配置 240

参考文献 242

第 1 章 Servlet

本章介绍 Servlet 及其相关的基础概念，包括 Servlet 容器、Servlet、ServletRequest、ServletResponse 等内容。本章案例是在浏览器窗口中显示简单文本。

通过学习本章内容，读者将可以：
- 了解 Servlet 及容器
- 创建 Servlet
- 向 Servlet 容器注册 Servlet

1.1 相关概念

Servlet 是 Web 开发的核心组件，而围绕 Servlet 的其他组件亦是不可或缺。这些组件各自提供不同的访问接口。

1.1.1 Servlet容器

Servlet 是实现了 jakarta.servlet.Servlet 接口且存在于 Web 服务端的对象，该对象由 Servlet 容器（又称 Servlet 引擎）动态装载并管理。Servlet 的作用是根据 Web 客户端的 HTTP 请求产生动态的 HTTP 应答；Servlet 并不直接与 Web 客户端交互，而是通过由 Servlet 容器实现的"请求/应答"接口（即 jakarta.servlet.ServletRequest 接口与 jakarta.servlet.ServletResponse 接口）与 Web 客户端进行交互。

Servlet 容器是 Web 服务器或者应用服务器的一个组件，它从两个方面对 Servlet 的功能进行支持。一方面，Servlet 容器负责装载和初始化 Servlet，并管理 Servlet 的生命周期，对于非分布式部署的 Servlet，Servlet 容器只保留 Servlet 单例；另一方面，Servlet 容器为 Servlet 提供接收 HTTP 请求及作出 HTTP 应答的网络服务：接收 Web 客户端的 HTTP 请求并基于 MIME（多用途互联网邮件扩展类型）对其解码以供 Servlet 使用，对 Servlet 产生的应答基于 MIME 编码并将其发送至 Web 客户端。

Servlet 容器对客户端请求的处理流程如下：

（1）客户端（如 Web 浏览器）访问 Web 服务器，发出 HTTP 请求。

（2）Web 服务器接收到请求，将其转交给 Servlet 容器。

（3）Servlet 容器根据 Servlet 配置信息决定调用哪一个 Servlet，调用时传入表示 HTTP 请求的 ServletRequest 对象以及表示 HTTP 应答的 ServletResponse 对象作为参数。

（4）Servlet 按照编程逻辑执行，可以从 ServletRequest 对象获取请求信息的数据，如客户端 IP 地址、HTTP 方法的参数等，处理后再通过 ServletResponse 对象把产生的数据（即应答信息）发送至客户端。

（5）当 Servlet 完成对请求的处理时，Servlet 容器会确保应答信息被真正地从缓冲区冲刷（flush），再把控制权归还给 Web 服务器。

图 1-1 展示了 Servlet 容器对信息的处理流程，连线代表信息传递方向，矩形框表示处理单元。

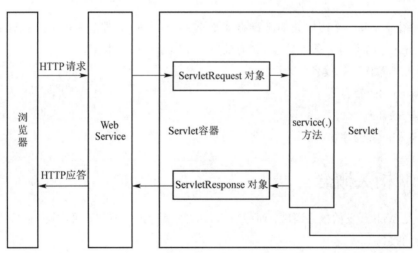

图1-1　Servlet容器对信息处理流程

1.1.2　ServletContext接口

ServletContext接口定义了Servlet对其所在Servlet容器的视图。Servlet 容器提供ServletContext接口的实现。通过ServletContext，一个Servlet 可以记录事件、获取资源的 URL 引用，以及设置或存储属性值以供 Servlet 容器中其他 Servlet 进行访问。

ServletContext 对应于 Web 服务器上某个应用程序 URL 的根。如果 Web 服务器 www.comany.com 的一个 Servlet 容器被配置成定位于 http://www.company.com/cwgl，此时/cwgl 称为容器路径（Context path），那么凡是请求路径以/cwgl 开头的请求均被转发至这个 ServletContext 对应的 Web 应用。ServletContext 与 Web 应用的关系如图 1-2 所示。

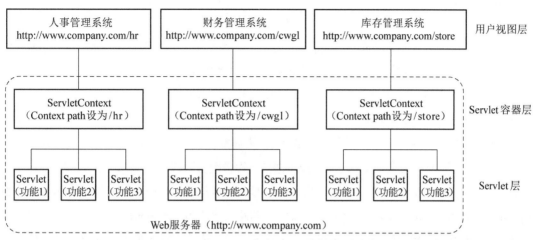

图1-2　ServletContext与Web应用的关系

每个 Web 应用对应一个 ServletContext 接口的实例对象，在分布式环境中，一个 Web 应用在每个 JVM 上有一个 ServletContext 实例。

当用编程方式配置 Tomcat 服务器的 ServletContext 时，可以调用 Tomcat 对象的 addContext(.) 方法。代码举例如下：

```
Tomcat tomcat = new Tomcat();
Context context = tomcat.addContext("/hr", System.getProperty("java.io.tmpdir"));
```

ServletContext 接口提供如下与配置 Servlet 有关的方法：

- ServletRegistration.Dynamic addServlet(String servletName, String className)
- ServletRegistration.Dynamic addServlet(String servletName, Servlet servlet)
- ServletRegistration.Dynamic addServlet(String servletName, Class <? extends Servlet> servletClass)
- <T extends Servlet> T createServlet(Class<T> clazz)
- ServletRegistration getServletRegistration(String servletName)
- Map<String, ? extends ServletRegistration> getServletRegistrations()

这些方法的参数 servletName 表示 Servlet 的逻辑名称，在向 ServletContext 注册 Servlet 时被指定；ServletRegistration.Dynamic 对象用于在注册后对 Servlet 进行设置。

向 Servlet 容器注册 Servlet 有三种方式：使用@WebServlet 注解，或者使用 XML 格式的部署描述符，或者用编程方式进行注册；@WebServlet 注解需要应用服务器的支持。

用编程方式向 ServletContext 注册 Servlet 的代码举例如下：

```
Servlet servlet = new HelloServlet();
ServletRegistration.Dynamic registration =
```

```
                        servletContext.addServlet ("hello", servlet);
registration.setLoadOnStartup(1);
registration.addMapping("/hello/*");
registration.setAsyncSupported(true);
```

配置 Servlet 时，registration.addMapping(.)方法的参数是一个表示路径模式（Path Pattern）的字符串，若浏览器某个请求的 URI 与此路径模式匹配，则该请求将由此 Servlet 处理。

路径模式的 URI 匹配规则，按照优先级从高到低顺序如下：

（1）优先作路径的精确匹配。

（2）以"/"开头且以"/*"结尾的模式串，*为通配符。

（3）以"*."开头的模式串用于扩展名匹配，*为通配符，如"*.jpg"。

（4）空模式串""是一个专门的 URI 模式，它精确地匹配到容器路径的根路径。例如，形如 http://host:port/<context-path>/ 的请求路径将匹配到路径模式为""的 Servlet。

（5）模式串"/"指明该 Servlet 为应用的默认 Servlet，凡是 URI 未匹配上述路径模式的请求将交由默认 Servlet 处理。

路径模式匹配的请求 URI 举例见表 1-1。

表 1-1 路径模式匹配举例

路径模式	可匹配到的请求 URI
/work/bar/*	❑ /work/bar/index.html ❑ /work/bar/user ❑ /work/bar/abc/walker
/job/*	❑ /job ❑ /job/index.html
/catalog	❑ /catalog
*.rar	❑ /catalog/apple.rar ❑ /index.rar
/	❑ /catalog/index.html（其他路径模式均不能匹配时）

1.1.3　Servlet接口

Servlet 接口是 Jakarta Servlet API 中最重要的一个。任何 Servlet 或者直接实现该接口，或者更多的是扩展一个已实现此接口的类。Jakarta Servlet API 中实现了 Servlet 接口的两个类是 GenericServlet 和 HttpServlet。多数情况下，开发者会扩展 HttpServlet 以实现自定义的功能。

Servlet 接口定义了一个 service(.)方法用于处理客户端的请求。客户端的每个请求被 Servlet 容器提交至某个 Servlet 对象的 service(.)方法处理。Servlet 对并发请求的处理要求 Web 应用的开发者考虑如何让 Servlet 对象的 service(.)方法不会因多线程并发执行引起数据访问冲突。

Servlet 接口的抽象子类 HttpServlet 增加了如下方法以帮助处理 HTTP 请求，这些方法在 service(.)方法内根据 HTTP 请求方法被自动调用：

- doGet(HttpServletRequest, HttpServletResponse)：处理 HTTP GET 请求。
- doPost(HttpServletRequest, HttpServletResponse)：处理 HTTP POST 请求。
- doPut(HttpServletRequest, HttpServletResponse)：处理 HTTP PUT 请求。
- doDelete(HttpServletRequest, HttpServletResponse)：处理 HTTP DELETE 请求。
- doHead(HttpServletRequest, HttpServletResponse)：处理 HTTP HEAD 请求。
- doOptions(HttpServletRequest, HttpServletResponse)：处理 HTTP OPTIONS 请求。
- doTrace(HttpServletRequest, HttpServletResponse)：处理 HTTP TRACE 请求。

当开发基于 HTTP 的 Servlet 时，通常只需关注 doGet(.)和 doPost(.)方法。

1.1.4 ServletRequest接口

在 HTTP 协议中，客户端的请求消息以 HTTP 消息头和 HTTP 消息体的形式从客户端传送至服务器。Servlet 容器把来自客户端的请求消息封装成 ServletRequest 对象。请求参数是请求消息的一部分，作为字符串从客户端传送至 Servlet 容器。Servlet 容器从 URI 查询字符串及 POST 方法所提交的数据中提取请求参数，这些参数以"参数名–参数值"配对的形式存储于 ServletRequest 对象内，一个参数名可以配对多个参数值。

ServletRequest 接口的以下方法用来访问请求参数：

- String getParameter(String paramName);
- Enumeration<String> getParameterNames();
- String[ ] getParameterValues(String paramName);
- Map<String, String[ ]> getParameterMap()。

其中，当参数相同时，getParameter(.)方法的返回值必定是 getParameterValues(.)方法所返回数组的第一个元素。

来自 URI 查询字符串的数据与 POST 所提交的数据都聚合于 ServletRequest 对象的请求参数集。查询字符串的数据出现于 POST 提交的数据之前。如果一个请求同时包含查询字符串 a=tiger 和 POST 数据 a=buffalo&a=panda，那么 ServletRequest 对象中的参数集将按照 a=(tiger, buffalo, panda)排列。

路径参数是 GET 请求的一部分，它们通过解析 getRequestURI()方法或者 getPathInfo()方法所返回的字符串可获得。

每个 ServletRequest 对象仅在 Servlet 的 service(.)方法内有效，或者在过滤器的 doFilter(.)方法内有效，除非 ServletRequest 对象调用了 startAsync(.)方法启动异步处理。当异步处理发生时，ServletRequest 对象会一直有效直至调用了 AsyncContext 的 complete()方法。

1.1.5 ServletResponse接口

ServletResponse接口的实现对象封装了所有从服务器发送至客户端的信息（即HTTP应答消息的消息头及消息体）及其操作。

Servlet容器可能因为效率原因缓存将发送至客户端的信息。当缓冲区填满时，容器会立即冲刷缓冲区（又称flush，即把缓冲区内容发送至客户端）。当第一个字节到达客户端时，此时称ServletResponse被提交。

不论Servlet使用ServletOutputStream还是使用Writer发送信息至客户端，Servlet都可以使用ServletResponse接口提供的以下方法操作缓冲区：

- ❑ int getBufferSize()
- ❑ void setBufferSize(int bufferSize)
- ❑ void flushBuffer()
- ❑ boolean isCommitted()
- ❑ void reset()
- ❑ void resetBuffer()

其中，setBufferSize(.)方法必须在使用ServletOutputStream或者Writer输出内容之前被调用；如果已经输出内容，则调用此方法会抛出异常。reset()与resetBuffer()都必须在应答尚未提交时调用，作用是删除缓冲区数据，否则会抛出异常；两者的区别在于reset()会额外删除调用此方法之前通过OutputStream或者Writer设置的应答消息头、状态码以及状态。

Servlet可以使用HttpServletResponse接口提供的以下方法设置HTTP应答的消息头：

- ❑ void setHeader(String, String)
- ❑ void addHeader(String, String)

前者用于替换同名消息头，后者用于追加同名消息头；两者都有新增消息头的作用。

消息头可能包含表示整数或者日期的数据，HttpServletResponse接口的以下方法允许Servlet用恰当类型的数据值作为参数设置消息头：

- ❑ void setIntHeader(String, int)
- ❑ void addIntHeader(String, int)
- ❑ void setDateHeader(String, long)
- ❑ void addDateHeader(String, long)

在ServletResponse被提交之前设置的消息头才能被输送到客户端，而在ServletResponse被提交之后设置的消息头将被Servlet容器忽略。

1.1.6 应答时字符编码

Servlet在应答时应该指明应答所采用的字符编码。Servlet所指明的字符编码方式保存在应答消息头Content-Type内，再发送至客户端。

ServletResponse对象的setCharacterEncoding(String)方法、setContentType(String)方法、setLocale(Locale)方法之一可用于指明应答的字符编码，但是在ServletResponse对象的getWriter()

方法被调用之后或者在应答被提交之后调用此方法则无效。如果 Servlet 在 ServletResponse 对象的 getWriter()方法调用之前或者应答被提交之前没有指明字符编码，那么将使用默认的 ISO-8859-1 编码，此时客户端收到的应答消息头内不含编码设置。

setContentType(.)方法只有在参数字符串中提供了 charset 属性值才具有指明字符编码的作用。以下是一个指明字符编码的例子：

```
resp.setContentType("text/html;charset=utf-8");
```

setLocale(.)方法只有在 setCharacterEncoding(.)方法及 setContentType(.)方法两者都没有指明字符编码时才具有指明字符编码的作用。

1.2 案例描述

用户在浏览器地址栏中输入地址 http://localhost:8080/?n=15&c=马，浏览器显示效果如图 1-3 所示。

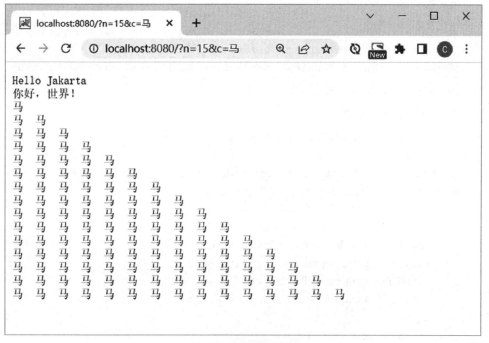

图1-3　案例运行界面

说明：地址栏中 n 与 c 是参数名，n 表示三角形的行数，c 表示三角形内的字符。

1.3 用Servlet实现

设计思路

浏览器用 GET 方法发出请求，用查询字符串传入参数，因此 Servlet 重载 doGet(.)方法，用 ServletRequest 的 getParameter(.)方法取得参数值，再调动自定义的功能类 Triangle 产生三角形图案的字符串，最后用 ServletResponse 将此字符串发送至浏览器。实现步骤如下：

第一步：新增 Servlet API 依赖。

代码 1-1：Servlet API 依赖

```xml
<dependency>
    <groupId>jakarta.servlet</groupId>
    <artifactId>jakarta.servlet-api</artifactId>
    <version>6.0.0</version>
</dependency>
```

第二步：在 TOMCAT 类的 service() 方法内在 Web 服务启动之前注册 ServletContext，并设置容器路径为""。

代码 1-2：注册 ServletContext

```java
// 向 tomcat 注册 ServletContext，并设置容器路径为""
Context context = tomcat.addContext("", System.getProperty("java.io.tmpdir"));
LifecycleListener lifecycleListener = (LifecycleListener)Class
                    .forName(tomcat.getHost().getConfigClass())
                    .getDeclaredConstructor().newInstance();
context.addLifecycleListener(lifecycleListener);
((StandardJarScanner) context.getJarScanner()).setScanManifest(true);
```

第三步：新增 HelloServlet 类。

代码 1-3：HelloServlet 类

```java
public class HelloServlet extends HttpServlet {
    @Override
    protected void doGet(HttpServletRequest req, HttpServletResponse resp)
            throws ServletException, IOException {
        //设置应答消息为简单文本类型
        resp.setContentType("text/plain");
        //此设置解决页面内中文显示为乱码的问题
        resp.setCharacterEncoding("utf-8");
        //读取 QueryString 中参数 n 的值
        String nParam = req.getParameter("n");
        int n = 5;
        if (nParam!=null){
            n = Integer.valueOf(nParam);
        }
        //读取 QueryString 中参数 c 的值
        String cParam = req.getParameter("c");
        Character character = '*';
        if (cParam!=null && cParam!=""){
            character = cParam.charAt(0);
        }
        //创建功能类 Triangle 实例
```

```
            Triangle triangle = new Triangle(n, character);
            //调用Triangle对象产生三角形图案字符串
            String msg = triangle.getShape();
            //向浏览器发送消息字符串
            resp.getWriter().println("Hello Jakarta\n 你好，世界！");
            //向浏览器发送三角形图案字符串
            resp.getWriter().println(msg);
        }
    }
```

Triangle 为功能类，根据参数 n 和 character 生成一个 n 行由 character 字符组成的三角形图案，假设 n 为 3，character 为星号*，则其 getShape()方法返回字符串"*\n**\n***\n"。

第四步：新增 ServletContainerInitializer 接口的实现类 ServletInitializer，重载 onStartup(.)方法，向 ServletContext 注册 HelloServlet。

代码 1-4：注册 HelloServlet

```
public void onStartup(Set<Class<?>> set, ServletContext servletContext)
throws ServletException {
    System.out.println("初始化各个Servlet,将其加入ServletContainer...");
    ServletRegistration.Dynamic registration;
    Servlet servlet;
    //注册HelloServlet
    servlet = new HelloServlet();
    registration = servletContext.addServlet("hello.servlet", servlet);
    registration.setLoadOnStartup(1);
    registration.addMapping("/");
    registration.setAsyncSupported(true);
}
```

Servlet 的路径模式设为"/"，表明此 Servlet 是所在 ServletContext 的默认处理器。

第五步：新建文件夹 resources\META-INF\services\，在其中新建名为 jakarta.servlet.ServletContainerInitializer 的文件，文件内容为上述第四步创建的类的全名 cxiao.sh.cn.init.ServletInitializer。

第六步：运行程序，之后在浏览器地址栏中输入 http://localhost:8080/?n=15&c=马，亦可手动调整参数 n 与 c 的值。

1.4 用Spring实现

设计思路

在 HelloController 的方法中使用@RequestParam 标注参数以获取 URL 的 QueryString 参数；在方法的@GetMapping 注解中设置应答编码方式；向 Web 服务器注册 ServletContext，在 WebApplicationInitializer 的实现类中向 ServletContext 注册 Spring 所定义的 DispatcherServlet。实现步骤如下：

第一步：新增 spring-webmvc 依赖。

代码 1-5：spring-webmvc 依赖

```xml
<dependency>
    <groupId>org.springframework</groupId>
    <artifactId>spring-webmvc</artifactId>
    <version>${spring.version}</version>
</dependency>
```

第二步：在 TOMCAT 类的 service()方法内 Web 服务启动之前注册 ServletContext，并设置容器路径为""。

代码 1-6：注册 ServletContext

```java
// 向 tomcat 注册 ServletContext，设置容器路径为""
Context context = tomcat.addContext("", System.getProperty("java.io.tmpdir"));
LifecycleListener lifecycleListener = (LifecycleListener)Class
                    .forName(tomcat.getHost().getConfigClass())
                    .getDeclaredConstructor().newInstance();
context.addLifecycleListener(lifecycleListener);
((StandardJarScanner)context.getJarScanner()).setScanManifest(true);
```

第三步：新增 HelloController 类，在@GetMapping 的 produces 属性中设置 charset=utf-8，用于解决页面中文乱码问题。

代码 1-7：HelloController 类

```java
@Controller
public class HelloController {
    @GetMapping(path ="/", produces = "text/plain;charset=utf-8")
    @ResponseBody
    public String hello(@RequestParam("n") @Nullable Integer nLayer,
                @RequestParam("c") @Nullable Character character){
        if (nLayer==null){
            nLayer = 5;
        }
        if(character==null){
            character = '*';
        }
        String msg = "Hello Spring\n 你好，世界！\n";
        Triangle triangle = new Triangle(nLayer, character);
        msg += triangle.getShape();
        return msg;
    }
}
```

第四步：新增 WebApplicationInitializer 接口的实现类 ServletInitializer，在重载方法 onStartup(.)中向 ServletContext 注册 Spring 预定义的 DispatcherServlet。

代码 1-8：ServletInitializer 类

```java
public class ServletInitializer implements WebApplicationInitializer {
    @Override
    public void onStartup(ServletContext servletContext) throws ServletException {
        //启动 Spring 上下文
        AnnotationConfigWebApplicationContext context =
                            new AnnotationConfigWebApplicationContext();
        context.register(MyApp.class);
        //创建 DispatcherServlet 对象
        Servlet servlet = new DispatcherServlet(context);
        //向 ServletContext 注册上述 DispatcherServlet
        ServletRegistration.Dynamic registration =
                            servletContext.addServlet("app", servlet);
        registration.setLoadOnStartup(1);
        registration.addMapping("/");
        registration.setAsyncSupported(true);
    }
}
```

第五步：运行程序，之后在浏览器地址栏中输入 http://localhost:8081/?n=15&c=马，亦可手动调整参数 n 与 c 的值。

1.5　用SpringBoot实现

设计思路

由于 SpringBoot 的自动配置特性，故无须显式设置应答编码，无须显式向 Web 服务器注册 ServletContext，也无须显式向 ServletContext 注册 DispatcherServlet。实现步骤如下：

第一步：新增 HelloController 类，注意此处@GetMapping 的 produces 属性不用指定 charset。

代码 1-9：HelloController 类

```java
@Controller
public class HelloController {
    @GetMapping(path ="/", produces = "text/plain")
    @ResponseBody
    public String hello(@RequestParam("n") @Nullable Integer nLayer,
                        @RequestParam("c") @Nullable Character character){
        if (nLayer==null){
            nLayer = 5;
        }
        if(character==null){
            character = '*';
        }
        String msg = "Hello SpringBoot\n你好，世界！\n";
```

```
            Triangle triangle = new Triangle(nLayer, character);
            msg += triangle.getShape();
            return msg;
        }
    }
```

第二步：运行程序，之后在浏览器地址栏中输入 http://localhost:8082/?n=15&c=马，亦可手动调整参数 n 与 c 的值。

1.6 小结

本章介绍了 Servlet 及其容器 ServletContext，ServletContext 由 Web 服务器提供，Web 应用开发就是创建不同功能的 Servlet 对象并将其添加至 Servlet 容器内。ServletRequest 与 ServletResponse 可分别看作 Servlet 在网络上的输入"设备"与输出"设备"。

1.7 习题

创建名为 Hello 的 Servlet，在浏览器地址栏中输入 http://localhost:8080，则浏览器以网页格式显示 Servlet 名称、请求的 URL 以及当前日期时间，页面效果如图 1-4 所示。

图1-4 浏览器显示效果

提示：应答的 ContentType 设置为 text/html，并且应答的消息体采用 HTML 格式文本。

第 2 章 过 滤 器

本章介绍过滤器。本章案例是在过滤器中修改请求参数的值以及设置应答的字符编码。通过学习本章内容，读者将可以：
- 阐述过滤器的概念
- 列举过滤器的功能
- 添加过滤器
- 通过包装器修改请求参数或应答内容

2.1 相关概念

过滤器可以分担 Servlet 处理之前或者处理之后的部分功能，这些功能原本可以直接写在 Servlet 的 service(.)方法内，但因为这些功能具有通用性，所以把它们放在过滤器中以便功能复用。

2.1.1 过滤器概念

过滤器（Filter）是配合 Servlet 使用的 Java 组件。Servlet 的作用是对接收到的请求进行处理然后做出应答，而过滤器的作用是在 Servlet 开始处理之前变换请求的消息头、消息体以及在 Servlet 处理结束之后变换应答的消息头、消息体。一个 Servlet 可以配置多个过滤器从而形成过滤器链，过滤器链的调用流程如图 2-1 所示。

过滤器本身不创建 ServletRequest 或者 ServletResponse 对象，而是修改或者调整 Servlet 所用的 ServletRequest 对象或者 ServletResponse 对象。

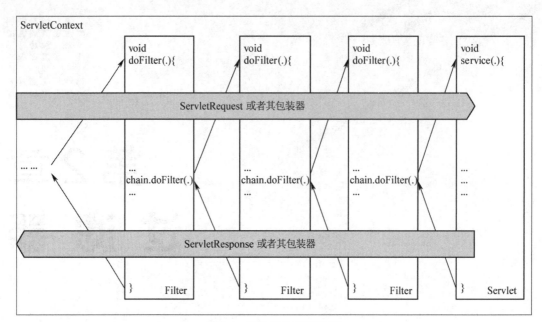

图2-1 过滤器的调用流程

开发时如果需要用到如下功能，就要考虑使用过滤器：
- 在 Servlet 调用之前对 ServletRequest 进行拦截
- 通过包装 ServletRequest 对象对请求消息头和消息体作自定义修改
- 通过包装 ServletResponse 对象对应答消息头和消息体作自定义修改
- 在 Servlet 调用之后对 ServletResponse 追加输出

一些特定功能通常放在过滤器中，如身份验证、日志审计、数据加密、输入缓存等。

开发者通过实现 jakarta.servlet.Filter 接口创建过滤器类，需要提供一个公开的无参构造器。在 ServletContainerInitializer 的实现类中向 ServletContext 注册过滤器，同时通过指明 Servlet 的逻辑名称把过滤器装配到某个特定 Servlet，或者通过指明 URI 路径模式字符串把过滤器装配到一组 Servlet 上。

在处理一个 HTTP 请求时，Servlet 的 service(.)方法与所有作用于此 Servlet 的过滤器运行于同一个线程。处理时 Serlvet 容器取出 Servlet 过滤器链中的第一个过滤器并且调用其 doFilter(.)方法，以准备传入 Servlet 的 ServletRequest 对象和 ServletResponse 对象，以及一个 FilterChain 对象作为参数。

一个过滤器的 doFilter(.)方法可以选用如下操作：

（1）用自定义的 HttpServletRequestWrapper 派生类来包装方法的参数 ServletRequest 对象，在生成的 ServletRequest 包装器对象中更改请求消息头或者请求消息体。

（2）用自定义的 HttpServletResponseWrapper 派生类来包装方法的参数 ServletResponse 对象，在生成的 ServletResponse 包装器对象中更改应答消息头或者应答消息体。

（3）调用 FilterChain 对象的 doFilter(.)方法，传入的参数则是当前 doFilter(.)被调用时所传入的 ServletRequest 对象和 ServletResponse 对象，或者是它们的包装器对象。

（4）如有必要，可以在 doFilter(.)方法内进行与 Servlet 内相同的操作，即用过滤器取代 Servlet。

（5）如果 doFilter(.)方法抛出异常，Servlet 容器将不再继续此过滤器链的处理。

过滤器"过滤"概念的核心是包装一个请求或者应答以便重写其行为来执行过滤任务。为了支持这种基于包装的过滤器，开发者必须保证不论是在调用 FilterChain 对象的 doFilter(.)方法时，还是在调用 RequestDispatcher 对象的 forward(.)方法或者 include(.)方法时，又或者是在调用 AsyncContext 对象的 dispatch(.)方法时，目标 Servlet 及其过滤器所接收的 ServletRequest 对象及 ServletResponse 对象必须是当前过滤器或者当前 Servlet 被传入的同一个 ServletRequest 对象及 ServletResponse 对象，或者其包装器对象。

2.1.2 包装器

在 Servlet 中通过 ServletRequest 的 getParameter(.)、getParameterNames()、getParameterValues(.)等方法获取请求参数，这些请求参数是无法直接修改的，也不能添加或者删除。

在过滤器中把 ServletRequest 包装至 HttpServletRequestWrapper 中并重写 ServletRequest 的相关方法，可以实现对请求参数的修改。

类似地，在过滤器中把 ServletResponse 包装至 HttpServletResponseWrapper 中并重写 ServletResponse 的相关方法，可以达到修改应答消息的效果。

2.1.3 配置过滤器

ServletContext 接口提供如下与配置过滤器有关的方法：

- FilterRegistration.Dynamic addFilter(String filterName, String className)
- FilterRegistration.Dynamic addFilter(String filterName, Filter filter)
- FilterRegistration.Dynamic addFilter(String filterName, Class <? extends Filter> filterClass)
- <T extends Filter> T createFilter(Class<T> clazz)
- FilterRegistration getFilterRegistration(String filterName)
- Map<String, ? extends FilterRegistration> getFilterRegistrations()

这些方法的参数 filterName 表示过滤器的逻辑名称，在向 ServletContext 注册过滤器时被指定；FilterRegistration.Dynamic 对象用于在注册后对过滤器进行设置。

向 Servlet 容器注册过滤器有三种方式：使用@WebFilter 注解，或者使用 XML 格式的部署描述符，或者用编程方式进行注册；@WebFilter 注解需要应用服务器的支持。

用编程方式向 ServletContext 注册过滤器的代码举例如下：

```
Filter filter = new CharacterEncodingFilter();
FilterRegistration filterRegistration = servletContext.addFilter("encodingFilter", filter);
filterRegistration.addMappingForUrlPatterns(
                    EnumSet.allOf(DispatcherType.class), true, "/*");
```

FilterRegistration 对象可以作为 servletContext.addFilter(.)方法的返回值，它提供两个方法用于设置过滤器：

- void addMappingForUrlPatterns(EnumSet<DispatcherType>, boolean, String...)
- void addMappingForServletNames(EnumSet<DispatcherType>, boolean, String...)

前者是基于路径模式及请求方式的过滤器匹配,此处路径模式与设置 Servlet 时的路径模式规则相同,但是当路径模式为"/"或者""时无效,不会触发过滤器;后者是基于 Servlet 名称及请求方式的过滤器匹配。

上述 FilterRegistration 对象的两个方法的第一个参数为请求方式,可以是以下 5 个值的任意组合:

- DispatcherType.REQUEST:对应于 sendRedirect(.)引发的或者直接来自客户端的请求
- DispatcherType.INCLUDE:对应于 include(.)引发的请求
- DispatcherType.FORWARD:对应于 forward(.)引发的请求
- DispatcherType.ASYNC:对应于 dispatch(.)引发的请求
- DispatcherType.ERROR:对应于 sendError(.)引发的或者因抛出异常引发的请求

其中,当此参数设为 null 时,其含义为 DispatcherType.REQUEST。

当请求引发 Servlet 执行时,会触发路径模式(或 Servlet 名称)及请求方式设置相符的过滤器的执行。例如,Servlet 的路径模式设为"/abc/*",过滤器的路径模式设为"*.doc"且请求方式设为 DispatcherType.REQUEST,则当浏览器请求 http://host:port/<context-path>/abc/xyz.doc 时,将引发此过滤器执行,若过滤器没有阻挡请求,则继续执行此 Servlet。根据基于包装的过滤器规则,此 Servlet 及此过滤器中的 req.getServletPath()相同,都是"/abc";而且两者的 req.getPathInfo()也相同,都是"/xyz.doc"。

2.2 案例描述

用户在浏览器地址栏中输入地址 http://localhost:8080/?n=4&c=马,浏览器显示效果如图 2-2 所示。

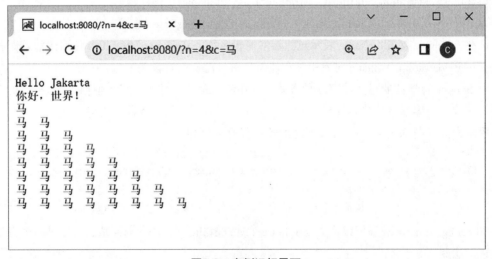

图2-2 案例运行界面

说明：地址栏中 n 与 c 是参数名，n 表示三角形行数的一半，如地址栏中 n=4，则浏览器窗口内显示的三角形有 8 行；c 表示三角形内的字符。

2.3 用Servlet实现

设计思路

在过滤器内设置应答的字符编码而不是在 Servlet 内设置应答的字符编码，这种方法可以适用于某一类访问而不限于某一个访问。新增两个过滤器，一个过滤器中设置应答的字符编码，这样避免在每个 Servlet 中分别设置应答的字符编码；另一个过滤器对 ServletRequest 进行包装，在包装器中修改参数 n 的值，将其值改为初始值的两倍。实现步骤如下：

第一步：删除 HelloServlet 的 doGet(.)方法内如下语句：

```
resp.setCharacterEncoding("utf-8");        //删除这行语句
```

第二步：新增 Filter 接口的实现类 CharacterEncodingFilter。

代码 2-1：CharacterEncodingFilter 类

```java
public class CharacterEncodingFilter implements Filter {
    @Override
    public void doFilter(ServletRequest servletRequest,
                         ServletResponse servletResponse,
                         FilterChain filterChain)
            throws IOException, ServletException {
        //解决网页的中文乱码问题，这个设置必须写在调用 doFilter(.)之前
        servletResponse.setCharacterEncoding("utf-8");
        filterChain.doFilter(servletRequest, servletResponse);
    }
}
```

第三步：新增 Filter 接口的实现类 TransformParameterFilter。

代码 2-2：TransformParameterFilter 类

```java
public class TransformParameterFilter implements Filter {
    @Override
    public void doFilter(ServletRequest servletRequest,
                         ServletResponse servletResponse,
                         FilterChain filterChain)
            throws IOException, ServletException {
        //构建 ServletRequest 对象的包装器
        HttpServletRequestWrapper requestWrapper =
            new HttpServletRequestWrapper((HttpServletRequest) servletRequest){
```

```java
        //重写getParameter(.)方法，因为在Servlet中调用了此方法
        @Override
        public String getParameter(String name) {
            //只修改参数n的值
            if(name.equals("n")){
                if(servletRequest.getParameter(name)!=null) {
                    int nValue =
                       Integer.parseInt(servletRequest.getParameter(name));
                    //将参数n的值改为原先值的2倍
                    return String.valueOf(2 * nValue);
                }
            }
            //其他参数值不变
            return servletRequest.getParameter(name);
        }
    };
    //把Wrapper作为参数调用过滤器链
    filterChain.doFilter(requestWrapper, servletResponse);
    }
}
```

考虑到 Servlet 可能需要用其他方法获取请求参数，这里更稳妥的做法是同时重载 getParameter(.)方法、getParameterValues(.)方法及 getParameterMap()方法。

第四步：在 ServletInitializer 的 onStartup(.)方法内向 ServletContext 注册上述两个过滤器。

代码 2-3：注册 CharacterEncodingFilter 过滤器及 TransformParameterFilter 过滤器

```
FilterRegistration filterRegistration;
Filter filter;
//增加此过滤器，用于处理页面显示中文的情况
filter = new CharacterEncodingFilter();
filterRegistration = servletContext.addFilter("encodingFilter", filter);
filterRegistration.addMappingForUrlPatterns(
            EnumSet.allOf(DispatcherType.class), true, "/*");
//注册TransformParameterFilter，用于修改参数n的值
filter = new TransformParameterFilter();
filterRegistration = servletContext.addFilter("tranformFilter", filter);
filterRegistration.addMappingForUrlPatterns(
            EnumSet.of(DispatcherType.REQUEST) , true, "/*");
```

第五步：运行程序，之后在浏览器地址栏中输入 http://localhost:8080/?n=4&c=马，亦可手动调整参数 n 与 c 的值。

2.4 用Spring实现

设计思路

在 Spring 中可以不采用在 Servlet 容器中添加过滤器的方式设置 ServletResponse 字符编码，而是在 Spring 容器中添加 StringHttpMessageConverter 对象完成编码转换；在 HelloController 的方法中带有@RequestParam 注解的参数值并不是从 ServletRequest.getParameter(.)方法获取的，而是从 ServletRequest.getParameterValues(.)方法获取的，因此在新增的 TransformParameterFilter 中用包装器重载 getParameterValues(.)方法。实现步骤如下：

第一步：修改 HelloController 的@GetMapping 中 produces = "text/plain"，即删除之前的 charset=utf-8。

第二步：新增 WebMvcConfigurer 接口的实现类 MvcConfig，重载方法 configureMessageConverters(.)。

代码 2-4：MvcConfig 类

```java
@Configuration
public class MvcConfig implements WebMvcConfigurer{
    @Override
    public void configureMessageConverters(
                    List<HttpMessageConverter<?>> converters) {
        //解决网页的中文乱码问题
        StringHttpMessageConverter stringConverter =
                new StringHttpMessageConverter(Charset.forName("UTF-8"));
        converters.add(stringConverter);
    }
}
```

第三步：新增 Filter 接口的实现类 TransformParameterFilter。

代码 2-5：TransformParameterFilter 类

```java
public class TransformParameterFilter implements Filter {
    @Override
    public void doFilter(ServletRequest servletRequest,
                    ServletResponse servletResponse,
                    FilterChain filterChain)
                    throws IOException, ServletException {
        //构建 ServletRequest 对象的 Wrapper
        HttpServletRequestWrapper requestWrapper =
            new HttpServletRequestWrapper((HttpServletRequest)servletRequest){
                //重写 getParameterValues(.)方法，Spring 中会调用此方法获取请求参数
                @Override
                public String[ ] getParameterValues(String name) {
                    String[ ] results = servletRequest.getParameterValues(name);
                    if(results == null || results.length <= 0) {
```

```
                    return null;
            }
            //只修改参数 n 的值
            if(name.equals("n")) {
                for(int i = 0; i < results.length; i++) {
                    int nValue = Integer.parseInt(results[i]);
                    //将参数 n 的值改为原先值的 2 倍
                    results[i] = String.valueOf(2 * nValue);
                }
            }
            return results;
        }
    };
    //把 Wrapper 作为参数调用过滤器链
    filterChain.doFilter(requestWrapper, servletResponse);
}
```

考虑到 Spring 可能需要用其他方法获取请求参数，这里更稳妥的做法是同时重载 getParameter(.)方法、getParameterValues(.)方法及 getParameterMap()方法。

第四步：在 ServletInitializer 的 onStartup(.)方法内向 ServletContext 注册上述过滤器。

代码 2-6：注册 TransformParameterFilter 过滤器

```
FilterRegistration filterRegistration;
Filter filter;
//注册 TransformParameterFilter 过滤器，用于修改参数 n 的值
filter = new TransformParameterFilter();
filterRegistration = servletContext.addFilter("tranformFilter", filter);
filterRegistration.addMappingForUrlPatterns(
                EnumSet.of(DispatcherType.REQUEST) , true, "/*");
```

第五步：运行程序，之后在浏览器地址栏中输入 http://localhost:8081/?n=4&c=马，亦可手动调整参数 n 与 c 的值。

2.5 用SpringBoot实现

设计思路

SpringBoot 的应答默认就是 UTF-8 编码，因此无须显式设置；与 Spring 的方法相同，参数 n 的值亦是在 TransformParameterFilter 中修改，但是 SpringBoot 另有注册过滤器的途径。实现步骤如下：

第一步：在文件夹 resources\内添加名为 application.properties 的文件，并在此文件内添加一行信息：server.servlet.encoding.charset=UTF-8。此项默认就是 UTF-8，可以不用显式设置。

第二步：新增 Filter 接口的实现类 TransformParameterFilter，与 2.4 节的第三步相同。

第三步：新增 WebMvcConfigurer 接口的实现类 MvcConfig，在其中创建类型为 FilterRegistrationBean 的 Bean，该 Bean 将被 SpringBoot 用于注册上述过滤器。

代码 2-7：创建用于注册 TransformParameterFilter 过滤器的 Bean

```java
@Configuration
public class MvcConfig implements WebMvcConfigurer{
    @Bean
    public FilterRegistrationBean<TransformParameterFilter>
            filterRegistrationBean(){
        FilterRegistrationBean<TransformParameterFilter> frBean =
                                     new FilterRegistrationBean<>();
        //设置要注册的 Filter
        frBean.setFilter(new TransformParameterFilter());
        //设置 Filter 的名称
        frBean.setName("tranformFilter");
        //设置 Filter 针对的请求方式
        frBean.setDispatcherTypes(EnumSet.of(DispatcherType.REQUEST));
        //设置 Filter 拦截哪些 URL
        frBean.addUrlPatterns("/*");
        //是否启用该 Filter，如果参数为 false，则不启用该 Filter
        frBean.setEnabled(true);
        //设置 Filter 执行顺序，值越小，越先执行
        frBean.setOrder(1);
        //设置 Filter 支持异步处理
        frBean.setAsyncSupported(true);
        return frBean;
    }
}
```

第四步：运行程序，之后在浏览器地址栏中输入 http://localhost:8082/?n=4&c=马，亦可手动调整参数 n 与 c 的值。

2.6 小结

本章介绍了过滤器的概念及其设置。过滤器的作用并不局限于字面的"过滤"功能，它更确切的功能是拦截与变换，变换必须在 ServletRequest 或者 ServletResponse 的包装器内进行；而 Spring、SpringBoot 并不需要创建过滤器就能实现应答的编码转换。

2.7 习题

使用过滤器在本章案例的基础上添加显示内容，浏览器显示效果如图 2-3 所示。说明：在开头及末尾各添加一行文字。

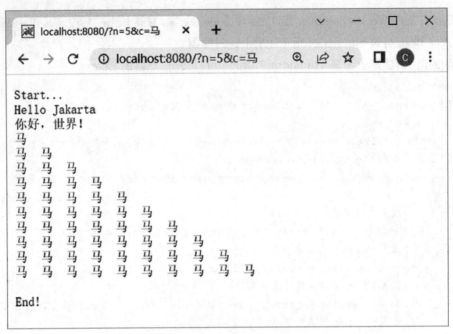

图2-3 浏览器显示效果

第 3 章
显示静态资源

本章介绍路径映射，把 URI 网址映射到 jar 包内或者文件系统内的静态文件。本章案例是在浏览器通过网址访问 jar 包内或文件系统内的静态文件。

通过学习本章内容，读者将可以：
- 从浏览器访问 jar 包内的静态资源文件
- 从浏览器访问文件系统内的静态资源文件

3.1 相关概念

在 HTML 网页中需要引用站点内的静态资源（如图片文件、CSS 文件等），这些资源可能包含在 Web 应用的 jar 包内，也可能位于服务器的文件系统内。

3.1.1 访问jar包资源

ServletContext 接口提供以下方法对 Web 应用 jar 包内的静态文档进行访问，如 CSS 文件、JavaScript 文件、Text 文件、JPEG 图片文件等：
- getResource(String)方法
- getResourceAsStream(String)方法

这两个方法接收以 "/" 开头的字符串作为参数，此参数指明静态资源相对于容器根的路径。这两个方法不能用于获取动态内容，例如，getResource("/index.jsp")将返回 JSP 源码而不是处理结果。另外，getResourcePaths(path)方法返回路径 path 下的全部资源列表。

3.1.2 路径匹配

用户在浏览器地址栏中输入的网址称为"请求 URL"（RequestURL），由"请求 URI"（RequestURI）与"查询字符串"（QueryString）用 "?" 连接而成，即

```
RequestURL = http://host:port/RequestURI + "?" + QueryString
```

而且

```
RequestURI = ContextPath + ServletPath + PathInfo
```

以下对每个部分进行说明：

❑ 容器路径（ContextPath）

用 HttpServletRequest 对象的 getContextPath()方法获取。

容器路径指在创建 Servlet 容器时指定的路径。容器路径或者是空字符串""，或者是以"/"开头且不以"/"结尾的字符串。

❑ Servlet 路径（ServletPath）

用 HttpServletRequest 对象的 getServletPath()方法获取。

Servlet 路径指在注册 Servlet 时参数"路径模式"所匹配到的请求路径。Servlet 路径或者是空字符串""（此为当路径模式为"/*"或者""时所匹配的内容），或者是以"/"开头且可能以"/"结尾的字符串。

❑ 路径信息（PathInfo）

用 HttpServletRequest 对象的 getPathInfo()方法获取。

路径信息指请求 URI 中排除容器路径及 Servlet 路径之后剩余的部分。路径信息或者是 null 或者是以"/"开头且可能以"/"结尾的字符串。

❑ 查询字符串（QueryString）

用 HttpServletRequest 对象的 getQueryString()方法获取。

查询字符串是请求 URL 中"?"后面的部分，形如 param1=value1&param2=value2 的字符串。它不作为请求 URI 的组成部分。

❑ 请求 URI

用 HttpServletRequest 对象的 getRequestURI()方法获取。

配置的路径模式与匹配的请求路径两者的匹配关系见表 3-1。

表 3-1 路径模式与请求路径匹配关系

优先级	Servlet 的路径模式 (Path Pattern)	匹配的 Servlet 路径 (ServletPath)	匹配的路径信息 (PathInfo)
1	/<servletPath>	"/<servletPath>"	null
2	/<servletPath>/*	"/<servletPath>" ❑ 当路径模式为"/*"时，匹配的 Servlet 路径为""。 ❑ 路径信息可能是 null 或者"/"或者以"/"开头的串	与"/*"通配对应
3	*.<扩展名>	与"*.<扩展名>"通配对应	null
4	""（空串）	"" 此路径模式只能匹配 http://host:port/<contextPath> 或者 http://host:port/<contextPath>/	"/"
5	/	请求 URI 减去容器路径 ❑ 默认的路径模式，匹配其余全部请求 URI。 ❑ Servlet 路径有可能以"/"结尾	null

举例说明见表 3-2。

表 3-2　路径模式与请求 URI 匹配举例（假设容器路径为/catalog）

Servlet 的路径模式	客户端的请求 URI	请求 URI 的各个组成部分
/boot/*	/catalog/boot/index.html	容器路径：/catalog Servlet 路径：/boot 路径信息：/index.html
/spring/*	/catalog/spring/implements/	容器路径：/catalog Servlet 路径：/spring 路径信息：/implements/
*.html	/catalog/help/feedback.html	容器路径：/catalog Servlet 路径：/help/feedback.html 路径信息：null

3.1.3　ServletResponse的关闭

当一个应答被关闭时，容器会立即将 ServletResponse 缓冲区的内容发送至客户端（即冲刷缓冲区）。以下事件表明 Servlet 已经满足了客户端的请求并且 ServletResponse 对象将要关闭：

- ❑ Servlet 的 service(.)方法及其出站过滤器执行完毕，除非已经启动异步模式。
- ❑ 通过 ServletResponse 输出的字节数已经达到了 ServletResponse 的 setContentLength(.)方法或者 setContentLengthLong(.)方法所设置的数量。
- ❑ ServletResponse 的 sendError(.)方法被调用。
- ❑ ServletResponse 的 sendRedirect(.)方法被调用。
- ❑ 异步模式下，AsyncContext 对象的 complete()方法被调用。

每个 ServletResponse 对象仅在 Servlet 的 service(.)方法内或者在过滤器的 doFilter(.)方法内有效，除非相关的 ServletRequest 对象启动了异步模式。在异步模式下，直到 AsyncContext 的 complete()方法被调用，ServletResponse 对象才会无效。

3.2　案例描述

项目的 resources\static\文件夹内有如图 3-1 所示文件；另外，操作系统的 d:\ImageOutside\文件夹内有 sun1.PNG、sun2.PNG、img\sun3.PNG 文件。

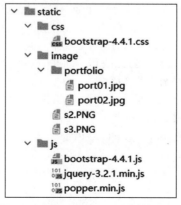

图3-1　resources\static\文件夹结构

分别在浏览器地址栏中输入以下 URL 地址，则浏览器窗口会显示对应的文件。

http://localhost:8080/css/bootstrap-4.4.1.css

http://localhost:8080/image/s2.png

http://localhost:8080/image/portfolio/port01.jpg

http://localhost:8080/js/bootstrap-4.4.1.js

http://localhost:8080/pic/sun1.png

http://localhost:8080/pic/img/sun3.png

其他静态文件的 URL 地址依此类推。

3.3 用Servlet实现

设计思路

根据浏览器所请求的静态资源 URI，定位到项目 jar 包内文件或者文件系统内文件，之后从文件读取字节序列，经由 ServletReponse 的输出流发送至浏览器。实现步骤如下：

第一步：新增 DefaultServlet 类。

代码 3-1：DefaultServlet 类

```java
public class DefaultServlet extends HttpServlet {
    //项目内静态资源的根路径
    private final String Static_Resource_PREFIX = "/static";
    //项目外静态资源的根路径
    private final String File_Resource_PREFIX = "D:/ImageOutside";
    @Override
    protected void doGet(HttpServletRequest req, HttpServletResponse resp)
                        throws ServletException, IOException {
        syncOutputResponse(req, resp);
    }
    private InputStream getInputStreamFromRequestedResource(HttpServletRequest req)
                                            throws IOException {
        InputStream is;
        //处理 /pic路径请求，这类 URI 要映射到文件系统的文件
        if (req.getServletPath().equals("/pic")){
            //获取项目外部静态资源的相对路径
            String fname = req.getPathInfo();
            //确定资源文件的完整路径
            String filePath = File_Resource_PREFIX + fname;
            //根据路径创建 File 对象
            File file = new File(filePath);
            if (!file.exists()){
                System.out.println(file.getCanonicalFile() + "不存在!!! ");
                return null;
            }
```

```
            //创建文件输入流
            is = new FileInputStream(file);
        }else {   //处理对项目内静态资源的访问请求
            //获取项目内部静态资源的相对路径
            String path = req.getServletPath() + req.getPathInfo();
            //根据资源的路径创建输入流
            is = this.getClass().getResourceAsStream(
                                Static_Resource_PREFIX + path);
        }
        return is;
    }
    private void syncOutputResponse(HttpServletRequest req,
                                    HttpServletResponse resp)
                                    throws IOException {
        //获取Response的输出流
        OutputStream os = resp.getOutputStream();
        //获取资源文件的输入流
        InputStream is = getInputStreamFromRequestedResource(req);
        //将输入流的字节序列转入输出流
        is.transferTo(os);
        //关闭输入流
        is.close();
        //冲刷输出流缓冲区
        os.flush();
    }
}
```

第二步：在 ServletInitializer 类的 onStartup(.)方法内注册 DefaultServlet 并配置其路径模式。

代码 3-2：注册 DefaultServlet

```
//注册 DefaultServlet
servlet = new DefaultServlet();
registration = servletContext.addServlet("default.servlet", servlet);
registration.setLoadOnStartup(2);
//前三个路径模式匹配项目内部静态文件，第四个路径模式匹配项目外部静态文件
registration.addMapping("/image/*", "/css/*", "/js/*", "/pic/*");
registration.setAsyncSupported(true);
```

第三步：运行程序，之后在浏览器地址栏中输入本章案例描述所列 URL 地址，查看浏览器窗口显示内容。

3.4 用Spring实现

设计思路

在 Spring 中无须创建 Servlet 来完成 URI 到文件的路径映射，只需要添加 ResourceHandler。

在 ResourceHandler 中进行路径映射，不但要设置 URI 与 jar 包内文件的映射，还要设置 URI 与外部文件的映射。实现步骤如下：

第一步：在 MvcConfig 类内重载 addResourceHandlers(.)方法。

代码 3-3：addResourceHandlers(.)方法

```
@Override
public void addResourceHandlers(ResourceHandlerRegistry registry) {
    //把静态资源 URI 映射到 jar 包内物理路径
    registry.addResourceHandler("/**")
                    .addResourceLocations("classpath:/static/");
    //把/pic/** URI 映射成文件系统物理路径
    registry.addResourceHandler("/pic/**")
                    .addResourceLocations("file:D:/ImageOutside/");
}
```

注意：spring MVC 的 URL 模式中存在 /** 模式，这一点与 Servlet 规范不同。

第二步：运行程序，之后在浏览器地址栏中输入本章案例描述所列 URL 地址（端口号改成 8081），查看浏览器窗口显示内容。

3.5 用SpringBoot实现

设计思路

在 SpringBoot 中亦是添加 ResourceHandler，在 ResourceHandler 中进行路径映射。根据 SpringBoot 的约定，静态资源默认放在 resources\static\文件夹内，故无须设置 URI 与 jar 包内文件的映射，只需要设置 URI 与外部文件的映射。实现步骤如下：

第一步：新增 WebMvcConfigurer 接口的实现类 MvcConfig，重载 addResourceHandlers(.)方法。

代码 3-4：MvcConfig 类

```
@Configuration
public class MvcConfig implements WebMvcConfigurer {
    @Override
    public void addResourceHandlers(ResourceHandlerRegistry registry) {
        //把/pic/** URI 映射成文件系统物理路径
        registry.addResourceHandler("/pic/**")
                        .addResourceLocations("file:D:/ImageOutside/");
    }
}
```

因为 jar 包内的静态资源已经放在 SpringBoot 约定的文件夹内，所以在 SpringBoot 中只需要配置 URI 到文件系统物理路径的映射。

第二步：运行程序，之后在浏览器地址栏中输入本章案例描述所列 URL 地址（端口号改成 8082），查看浏览器窗口显示内容。

3.6 小结

本章介绍如何把 URI 映射到资源文件，这些资源文件在 HTML 网页内将以 URI 的方式被引用。Servlet 是通过应答输出流将资源文件的内容发送至客户端，而 Spring 及 SpringBoot 只需要设置映射关系。

3.7 习题

在本章案例的基础上显示网页 main.html，浏览器显示效果如图 3-2 所示。

图3-2 浏览器显示效果

说明：素材文件 main.html 位于"源码\chapter03\习题附件"文件夹内。要求在浏览器地址栏中输入 http://localhost:8080/image/main.html 或者 http://localhost:8080/pic/main.html 均可显示图示效果。

第 4 章
显示模板网页

本章介绍监听器与Thymeleaf模板引擎。本章案例是在浏览器通过网址访问HTML模板网页。通过学习本章内容，读者将可以：
- 显示 HTML 模板网页
- 使用监听器

4.1 相关概念

Thymeleaf 与 JSP 都是网页模板引擎，用于根据网页模板生成 HTML 网页。虽然 Jakarta Servlet 规范使用的是 JSP，但在实际开发中，选用 Thymeleaf 的系统日益增多。这得益于 Spring 及 SpringBoot 使用的是 Thymeleaf 模板引擎，而且 Thymeleaf 更便于前后端分离。

4.1.1 Thymeleaf

Thymeleaf 是服务端的模板引擎，它具有如下特点：
- Thymeleaf 语法接近 HTML
- Thymeleaf 支持 HTML 5 标准
- Thymeleaf 使用 OGNL 表达式访问对象
- Thymeleaf 扩展性较好

Thymeleaf 标准表达式见表 4-1。

表 4-1　Thymeleaf 标准表达式分类表

简单表达式	
变量表达式	${...}
选择变量表达式	*{...}
消息表达式	#{...}
链接表达式	@{...}
片段表达式	~{...}

续表

字面量	
文本	'one text', 'Another one!', ...
数值	0, 34, 3.0, 12.3, ...
布尔	true, false
空值	null
文字标记	one, sometext, main, ...
文本操作	
字符串连接	+
文本替换	|The name is ${name}|
算术运算	
二元运算符	+, -, *, /, %
负号（一元运算符）	-
布尔运算	
二元运算符	and, or（或者&&, ||）
布尔否定（一元运算符）	not（或者!）
比较和相等	
比较	>, <, >=, <=（或者 gt, lt, ge, le）
相等判断	==, !=（或者 eq, ne）
条件运算符	
IF-THEN	(if) ? (then)
IF-THEN-ELSE	(if) ? (then) : (else)
默认	(value) ?: (defaultvalue)

下面这个示例涵盖了上述大部分表达式：

```
'User is ' + (${user.isAdmin()} ? 'Administrator' : (${user.type}?:'Unknown'))
```

Thymeleaf 常用属性见表 4-2。

表 4-2　Thymeleaf 常用属性表

Thymeleaf 属性	功　　能
th:value	用来设置表单控件的值。例如： `<input th:value="${user.account}" ...>`
th:name	用来设置表单控件名称。例如： `<input th:name="account" ...>`
th:text	如果变量有值，则替换标签的默认文本，否则展示标签的默认文本。例如： `<p th:text="${msg}">默认文本</p>`
th:href	用来指向某个 URL，常与@{...}搭配使用。例如： `<a th:href="@{/user}">用户管理</a>`
th:action	用来配置表单 form 的请求路径，常与@{...}搭配使用。例如： `<form th:action="@{/login}" ...>`
th:src	用来指向文件的 URL，常与@{...}搭配使用。例如： `<img th:src="@{${imageInfo.imageUri}}" />`
th:if	条件判断，如果判断为真就显示所在标签或者执行相关操作。例如： `<div th:if="${errors!=null and errors.get('languages')!=null}" ...>`

续表

Thymeleaf 属性	功 能
th:onclick	绑定事件。例如： <a href="javascript:;" th:onclick="falling('resume')">继续</a>

HTML 文件必须以如下方式引入 Thymeleaf 的命名空间才能使用表 4-2 所列属性。
`<html xmlns:th="http://www.thymeleaf.org">`

4.1.2 ServletContext命名属性

Servlet 或者过滤器可以通过指定的名称把一个对象绑定到 ServletContext 使之成为 ServletContext 的命名属性。任何绑定到 ServletContext 的对象也可以被其他 Servlet 及过滤器访问和使用。以下是 ServletContext 提供的相关方法：

- setAttribute(String, Object)方法
- getAttribute(String)方法
- getAttributeNames()方法
- removeAttribute(String)方法

4.1.3 事件与监听器

一个监听器就是一组回调函数，由特定的事件引发其执行。Web 应用的事件监听机制让开发者对 ServletContext、HttpSession 和 ServletRequest 的生命周期有更强的控制。事件监听器是实现了一个或多个 EventListener 接口的类，它们在 Web 应用启动时初始化并向 Servlet 容器注册。

事件监听器在 ServletContext、HttpSession 及 ServletRequest 对象的状态发生变化时接收到事件通知并进行相应的处理。ServletContext 监听器用来管理应用级别的资源或状态；HttpSession 监听器用来管理与同一个浏览器的一系列请求有关的资源或状态；ServletRequest 监听器用来管理跨越 ServletRequest 生命周期的资源或状态。异步监听器用来管理异步事件，比如异步处理的超时或者结束。

可以有多个监听器同时对一种事件类型进行监听，开发者可以指定当事件发生时 Servlet 容器执行监听器的顺序。

事件类型及其对应的监听器接口见表 4-3 至表 4-5。

表 4-3 ServletContext 的事件类型及其监听器接口

ServletContext 的 事件类型	描 述	监听器接口
生命周期	❑ ServletContext 刚被创建 ❑ ServletContext 将要关闭	ServletContextListener
属性变化	❑ 添加 ServletContext 属性 ❑ 删除 ServletContext 属性 ❑ 替换 ServletContext 属性	ServletContextAttributeListener

表 4-4　HttpSession 的事件类型及其监听器接口

HttpSession 的事件类型	描述	监听器接口
生命周期	❑ HttpSession 刚被创建 ❑ HttpSession 作废 ❑ HttpSession 超时	HttpSessionListener
属性变化	❑ 添加 HttpSession 属性 ❑ 删除 HttpSession 属性 ❑ 替换 HttpSession 属性	HttpSessionAttributeListener
ID 变化	❑ HttpSession 的 ID 被改变	HttpSessionIdListener
Session 迁移	❑ HttpSession 活化 ❑ HttpSession 钝化	HttpSessionActivationListener
对象绑定	❑ 对象被绑定到 HttpSession ❑ 对象从 HttpSession 解除绑定	HttpSessionBindingListener

表 4-5　ServletRequest 的事件类型及其监听器接口

ServletRequest 的事件类型	描述	监听器接口
生命周期	❑ ServletRequest 刚被创建 ❑ ServletRequest 将要销毁	ServletRequestListener
属性变化	❑ 添加 ServletRequest 属性 ❑ 删除 ServletRequest 属性 ❑ 替换 ServletRequest 属性	ServletRequestAttributeListener
异步事件	❑ 异步处理超时 ❑ 异步处理连接终止 ❑ 异步处理结束	AsyncListener

举例说明监听器的使用。假设一个使用数据库且包含多个 Servlet 的 Web 应用，开发者提供一个 ServletContext 监听器按如下方式管理数据库连接：

（1）当应用启动时，监听器被调用，它登录数据库并且在 ServletContext 中存储数据库连接。

（2）Servlet 根据需要从 ServletContext 中取得数据库连接并使用它。

（3）当 Web 服务器关闭时，监听器接收到通知，然后关闭在 ServletContext 中存储的数据库连接。

4.1.4　注册监听器

新建的监听器类必须实现以下接口的一个或多个：

❑ jakarta.servlet.ServletContextListener

❑ jakarta.servlet.ServletContextAttributeListener

❑ jakarta.servlet.ServletRequestListener

❑ jakarta.servlet.ServletRequestAttributeListener

❑ jakarta.servlet.http.HttpSessionListener

- jakarta.servlet.http.HttpSessionAttributeListener
- jakarta.servlet.http.HttpSessionIdListener

向Servlet容器注册监听器有三种方式：使用@WebListener注解，或者使用XML格式的部署描述符，或者用编程方式进行注册；@WebListener注解需要应用服务器的支持。

以编程方式向ServletContxt注册监听器的函数如下：

- void addListener(String className)
- <T extends EventListener> void addListener(T t)
- void addListener(Class <? extends EventListener> listenerClass)
- <T extends EventListener> void createListener(Class<T> clazz)

当事件发生时，监听器按照它们注册的顺序依次被通知；但也有例外，比如HttpSessionListener.sessionDestroyed(.)按照注册相反的顺序被调用。

4.2 案例描述

在浏览器地址栏中输入网址 http://localhost:8080/index，则页面效果如图4-1所示。

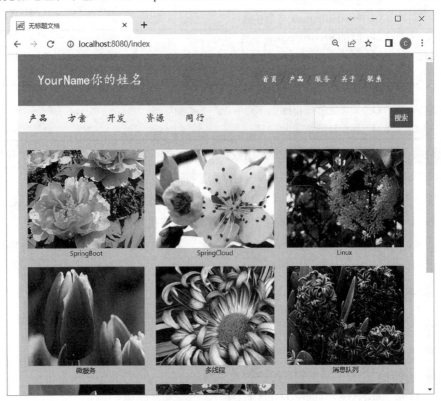

图4-1 案例运行界面

4.3 用Servlet实现

设计思路

在 Servlet 容器初始化时创建 Thymeleaf 引擎并保存到 Servlet 容器中。在需要输出网页至客户端时，先从 Servlet 容器取出所保存的 Thymeleaf 引擎，再调用此引擎的方法完成网页的输出。实现步骤如下：

第一步：在 resources\ 内创建图 4-2 所示的文件夹结构，并添加相应的文件。

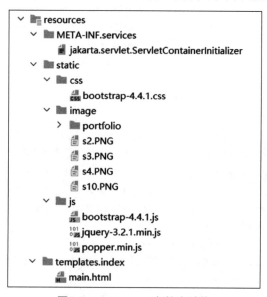

图4-2　resources\文件夹结构

第二步：在 pom.xml 中添加 Thymeleaf 依赖。

代码 4-1：Thymeleaf 依赖

```xml
<dependency>
    <groupId>org.thymeleaf</groupId>
    <artifactId>thymeleaf</artifactId>
    <version>3.1.1.RELEASE</version>
</dependency>
<dependency>
    <groupId>org.thymeleaf.extras</groupId>
    <artifactId>thymeleaf-extras-java8time</artifactId>
    <version>3.0.4.RELEASE</version>
</dependency>
```

第三步：创建工具类 ThymeleafUtil，使用 Thymeleaf API 提供 HTML 页面显示功能。

代码 4-2：ThymeleafUtil 类

```java
public class ThymeleafUtil {
    //ServletContext 存放模板引擎的属性名称
```

```java
        private static final String TEMPLATE_ENGINE_ATTR =
                    "cn.shanghai.cxiao.thymeleaf.TemplateEngineInstance";
    static public void outputHTML(HttpServletRequest req,
                                   HttpServletResponse resp,
                                   Map<String, Object> map,
                                   String templateName)
                                   throws IOException {
        ThymeleafUtil.synchronizedOutputHTML(req, resp, map, templateName);
    }
    static private void synchronizedOutputHTML(HttpServletRequest req,
                                   HttpServletResponse resp,
                                   Map<String, Object> map,
                                   String templateName)
                                   throws IOException {
        //从ServletContext的属性取得模板引擎
        ITemplateEngine templateEngine = (ITemplateEngine) req.getServletContext()
                .getAttribute(ThymeleafUtil.TEMPLATE_ENGINE_ATTR);
        //使用Thymeleaf构建当前ServletContext的WebExchange
        IWebExchange webExchange = JakartaServletWebApplication
                                   .buildApplication(req.getServletContext())
                                   .buildExchange(req, resp);
        //创建Thymeleaf的WebContext对象
        WebContext context = new WebContext(webExchange, req.getLocale());
        //参数map用于存放要填入到网页模板内的数据,通过键名获取
        if(map!=null) {
            for(String key : map.keySet()) {
                //把需要在HTML模板内展示的数据添加至WebContext
                context.setVariable(key, map.get(key));
            }
        }
        //使用模板引擎,把由HTML模板所生成的网页通过Response发送
        templateEngine.process(templateName, context, resp.getWriter());
    }
    //该方法只在ServletContextListener中调用一次
    static public void initializeTemplateEngine(ServletContext servletContext){
        //创建模板引擎
        ITemplateEngine templateEngine = ThymeleafUtil.templateEngine();
        //此模板引擎保存到ServletContext中,方便以后检索并使用
        servletContext.setAttribute(TEMPLATE_ENGINE_ATTR,templateEngine);
    }
    static private ITemplateEngine templateEngine() {
        ClassLoaderTemplateResolver templateResolver =
                                   ThymeleafUtil.templateResolver();
        TemplateEngine templateEngine = new TemplateEngine();
```

```java
        templateEngine.setTemplateResolver(templateResolver);
        //用于支持在网页模板中显示LocalDate等Java8数据类型
        templateEngine.addDialect(new Java8TimeDialect());
        return templateEngine;
    }
    static private ClassLoaderTemplateResolver templateResolver() {
        ClassLoaderTemplateResolver templateResolver =
                            new ClassLoaderTemplateResolver();
        //HTML是默认的模板模式，可以不用设置
        templateResolver.setTemplateMode(TemplateMode.HTML);
        //这个设置会把模板名称如"home"转换至"/templates/home.html"
        templateResolver.setPrefix("/templates/");
        templateResolver.setSuffix(".html");
        templateResolver.setCharacterEncoding(StandardCharsets.UTF_8.name());
        //取消模板的缓存，当模板被修改时会自动更新
        templateResolver.setCacheable(false);
        templateResolver.setOrder(1);
        return templateResolver;
    }
}
```

第四步：定义 ServletContext 监听器 GlobalObjectSetup，在 Servlet 容器初始化时完成 Thymeleaf 模板引擎的初始化。

代码 4-3：GlobalObjectSetup 类

```java
public class GlobalObjectSetup implements ServletContextListener {
    @Override
    public void contextInitialized(ServletContextEvent sce) {
        ServletContext servletContext = sce.getServletContext();
        //创建全局的 TemplateEngine 对象并存入 ServletContext
        ThymeleafUtil.initializeTemplateEngine(servletContext);
        //此处创建其他全局对象
        //...（暂无）
    }
}
```

第五步：在 ServletInitializer 的 onStartup(.)方法内向 ServletContext 注册监听器 GlobalObjectSetup。

代码 4-4：注册 GlobalObjectSetup 监听器

```java
//注册 GlobalObjectSetup 监听器
ServletContextListener servletContextListener;
servletContextListener = new GlobalObjectSetup();
servletContext.addListener(servletContextListener);
```

第六步：新增 IndexServlet 类。

代码 4-5：IndexServlet 类

```java
public class IndexServlet extends HttpServlet {
```

```
    @Override
    protected void doGet(HttpServletRequest req, HttpServletResponse resp)
                                    throws ServletException, IOException {
        String templateName = "index/main";
        //目前没有需要在网页中显示的变量值，所以第 3 个参数设为 null
        ThymeleafUtil.outputHTML(req, resp, null, templateName);
    }
}
```

第七步：在 ServletInitializer 的 onStartup(.)方法内向 ServletContext 注册 IndexServlet。

代码 4-6：注册 IndexServlet

```
//注册 IndexServlet
servlet = new IndexServlet();
registration = servletContext.addServlet("index.servlet", servlet);
registration.setLoadOnStartup(3);
registration.addMapping("/index");
registration.setAsyncSupported(true);
```

第八步：运行程序，之后在浏览器地址栏中输入 http://localhost:8080/index，查看显示结果。

4.4 用Spring实现

设计思路

在 Spring 容器中创建 ThymeleafViewResolver Bean，IndexController 类的方法返回的是网页模板名称，Spring 会自动调用 ThymeleafViewResolver 的方法将网页模板转换成 HTML 网页并显示。实现步骤如下：

第一步：在 resources\ 内创建图 4-3 所示的文件夹结构，并添加相应的文件。

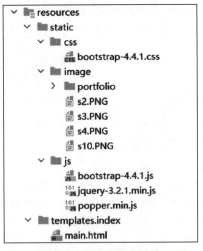

图4-3　新增文件夹结构

第二步：在模块的 pom.xml 文件内添加 Thymeleaf 依赖。

代码 4-7：Spring 的 Thymeleaf 依赖

```xml
<!--Thymeleaf 依赖-->
<dependency>
    <groupId>org.thymeleaf</groupId>
    <artifactId>thymeleaf-spring6</artifactId>
    <version>3.1.1.RELEASE</version>
</dependency>
<dependency>
    <groupId>org.thymeleaf.extras</groupId>
    <artifactId>thymeleaf-extras-java8time</artifactId>
    <version>3.0.2.RELEASE</version>
</dependency>
```

第三步：在 MvcConfig 中添加与 Thymeleaf 相关的 Bean。

代码 4-8：Thymeleaf 相关的 Bean

```java
@Bean
public ViewResolver viewResolver(SpringTemplateEngine templateEngine) {
    ThymeleafViewResolver viewResolver = new ThymeleafViewResolver();
    viewResolver.setTemplateEngine(templateEngine);
    viewResolver.setCharacterEncoding("UTF-8");
    return viewResolver;
}
@Bean
public SpringTemplateEngine templateEngine(ITemplateResolver templateResolver) {
    SpringTemplateEngine templateEngine = new SpringTemplateEngine();
    templateEngine.setTemplateResolver(templateResolver);
    return templateEngine;
}
@Bean
public ITemplateResolver templateResolver() {
    SpringResourceTemplateResolver templateResolver =
                             new SpringResourceTemplateResolver();
    //此处用 "classpath:templates/" 也行
    templateResolver.setPrefix("classpath:/templates/");
    templateResolver.setSuffix(".html");
    templateResolver.setTemplateMode(TemplateMode.HTML);
    templateResolver.setCharacterEncoding("UTF-8");
    templateResolver.setCacheable(false);
    templateResolver.setOrder(1);
    return templateResolver;
}
```

第四步：创建 IndexController 类。

代码 4-9：IndexController 类

```
@Controller
public class IndexController {
    @GetMapping("/index")
    public String index(){
        return "index/main";
    }
}
```

第五步：运行程序，之后在浏览器地址栏中输入 http://localhost:8081/index，查看显示结果。

4.5　用SpringBoot实现

设计思路

SpringBoot 会根据 Thymeleaf 依赖自动创建 ThymeleafViewResolver Bean，因此在 SpringBoot 中无须显式创建 ThymeleafViewResolver Bean。只需通过 IndexController 类的方法返回网页模板名称。实现步骤如下：

第一步：在 resources\ 内创建图 4-4 所示的文件夹结构，并添加相应的文件。

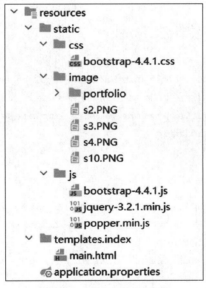

图4-4　新增文件夹结构

第二步：在模块的 pom.xml 文件内添加 Thymeleaf 依赖。

代码 4-10：SpringBoot 的 Thymeleaf 依赖

```
<dependency>
    <groupId>org.springframework.boot</groupId>
    <artifactId>spring-boot-starter-thymeleaf</artifactId>
    <version>${springboot.version}</version>
</dependency>
```

第三步：创建 IndexController 类。

代码 4-11：IndexController 类

```
@Controller
public class IndexController {
    @GetMapping("/index")
    public String index(){
        return "index/main";
    }
}
```

第四步：运行程序，之后在浏览器地址栏中输入 http://localhost:8082/index，查看显示结果。SpringBoot 自动完成了很多配置工作。

4.6 小结

本章介绍如何用 Thymeleaf 模板引擎将网页模板生成 HTML 网页。浏览器所请求的 URL 被映射到 Servlet，Servlet 动态生成 HTML 网页答复给浏览器，因此 HTML 网页不属于静态资源，它所用到的网页模板本身也不需要进行 URI 路径配置。由于模板引擎是一个共享对象，因此通过 ServletContextListener 在 ServletContext 初始化时将其创建并保存于 ServletContext 的命名属性中。

4.7 习题

在本章案例的基础上添加如下功能：在浏览器地址栏中输入 http://localhost:8080/index?name=张三，浏览器显示效果如图 4-5 所示。

图4-5　浏览器显示效果

提示：需要在 main.html 的 \<html\> 标签内添加属性 xmlns:th="http://www.thymeleaf.org"。

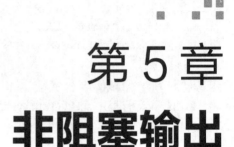

第 5 章
非阻塞输出

本章介绍 Servlet 的非阻塞输出。本章案例是以非阻塞的方式输出静态文件及 HTML 网页至浏览器。

通过学习本章内容，读者将可以：
- 为静态资源增加非阻塞输出功能
- 为动态网页增加非阻塞输出功能

5.1 相关概念

同步处理与异步处理的区别在于是否为多线程，非阻塞输出是异步处理的一种方式。非阻塞输出是采用监听器模式，在输出流状态发生变化时自动引发回调函数的异步执行。

5.1.1 异步输出

Servlet 容器对请求的非阻塞处理有助于满足 Web 服务器日益增长的可扩展性需求，有助于提高 Servlet 容器可并发处理的连接数。Servlet 容器的非阻塞输入允许程序在信道有数据到达时才读取数据，这避免了等待数据到达；而 Servlet 容器的非阻塞输出允许程序在信道可以发送数据时再写出数据，这样避免了发送阻塞。图 5-1 所示为同步处理流程，图 5-2 所示为异步处理流程。

非阻塞输入/输出只有在 Servlet 和过滤器可以进行异步处理时才有效，否则在调用 ServletInputStream.setReadListener(.) 或者 ServletOutputStream.setWriteListener(.) 时会抛出 IllegalStateException 异常，因此需要先切换至异步模式。

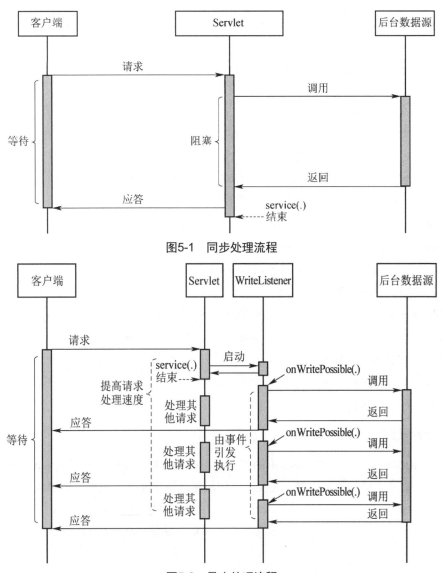

图5-1 同步处理流程

图5-2 异步处理流程

5.1.2 AsyncContext

对请求的异步处理允许对慢速资源的函数调用立即返回不被阻塞以便 Servlet 执行其他任务。在启动异步处理模式后，可以让另一个线程或者 WriteListener 的回调过程产生应答再调用 AsyncContext 的 complete()方法，也可以使用 AsyncContext 的 dispatch(.)方法将请求移交至另一个 Servlet 进行处理。

Servlet 通过调用其参数 ServletRequest 对象的 startAsync(.)方法开启异步模式 AsyncContext，再通过调用 AsyncContext 实例的 start(Runnable)方法启动另一个线程的异步处理，而原来的线程继续执行，最终以进入过滤器链相反的顺序退出过滤器链；退出后 ServletResponse 并不关闭，因此在异步线程中可以使用 ServletResponse 发送应答，直到 AysncContext 的 complete()方法被调用时 ServletResponse 才关闭。只有 asyncSupported 设为 true 的 Servlet 才能调用上述 startAsync(.) 方法，否则将抛出异常。

调用 ServletRequest 的以下方法可获得 AsyncContext 接口的实例：

- ❑ AsyncContext startAsync(ServletRequest req, ServletResponse resp)
- ❑ AsyncContext startAsync()

第一个方法的参数 req 和 resp 必须与调用者 Servlet 的 service(.)方法的参数是同一个对象或者是其包装器对象。这两个方法的调用必须满足三个条件：ServletRequest 位于支持异步操作的 Servlet 内、ServletResponse 尚未关闭、不能在调用 dispatch(.)后再次调用。

AsyncContext 提供以下方法用于支持异步操作：

- public ServletRequest getRequest()
- public ServletResponse getResponse()
- public void setTimeout(long timeoutMilliseconds)
- public long getTimeout()
- public void addListener(AsyncListener listener, ServletRequest req, ServletResponse res)
- public <T extends AsyncListener> createListener(Class<T> clazz)
- public void addListener(AsyncListener)
- public void dispatch(String path)
- public void dispatch()

5.1.3　WriteListener

ServletOutputStream 的 setWriteListener(WriteListener)方法用于向其注册 WriteListener。注册 WriteListener 后将启动非阻塞输出，这时再使用阻塞式输出是不合法的。非阻塞输出意味着当 ServletOutputStream 上可以写出数据时，Servlet 容器会调用 WriteListener 的回调方法，程序必须在回调方法里进行输出。WriteListener 提供以下回调方法供 Servlet 容器在对应的事件发生时调用：

- ❑ void onWritePossible()
- ❑ void onError(Throwable t)

第一个函数的调用发生在写操作从不可能变成可能的转折点（包括注册后第一次可以写出数据时），即 ServletOutputStream 的 isReay()方法值从 false 向 true 转变的瞬间；ServletOutputStream 输出应答时发生错误将导致第二个函数被 Servlet 容器调用。

5.2　案例描述

在浏览器地址栏中输入网址 http://localhost:8080/index，页面显示效果如图 5-3 所示。

图5-3　案例运行界面

说明：HTML 网页本身采用非阻塞输出，而且网页内所引用的静态资源亦采用非阻塞输出。

5.3　用Servlet实现

设计思路

新建 WriteListenerForResponse 类，该类对其构造参数 ServletResponse 进行监听，当 ServletResponse 可以输出时 Servlet 容器会自动执行此 WriteListenerForResponse 的回调函数进行输出。分别用输出静态资源的 ServletResponse 及输出 HTML 网页的 ServletResponse 作为构造参数创建 WriteListenerForResponse 实例，利用监听特性对静态资源及 HTML 网页进行非阻塞输出。实现步骤如下：

第一步：新增 WriteListenerForResponse 类。

代码 5-1：WriteListenerForResponse 类

```
public class WriteListenerForResponse implements WriteListener {
    private AsyncContext asct;
    InputStream is;
    public WriteListenerForResponse(InputStream is,
                                    HttpServletRequest req,
                                    HttpServletResponse resp){
        this.is = is;
        asct = req.startAsync(req, resp);
        asct.setTimeout(-1);
    }
```

```java
    public void setup() throws IOException {
        asct.getResponse().getOutputStream().setWriteListener(this);
    }
    @Override
    public void onWritePossible() throws IOException {
        ServletOutputStream os = asct.getResponse().getOutputStream();
        byte[ ] buffer = new byte[1024*1024];
        int n;
        while(os.isReady()) {
            if((n = is.read(buffer, 0, buffer.length)) > 0) {
                os.write(buffer, 0, n);
            }else{       //输入源已空，故关闭is，及终结AsyncContext
                is.close();
                asct.complete();
            }
        }
    }
    @Override
    public void onError(Throwable throwable) {
        asct.complete();
        try {
            is.close();
        }catch (Exception ex){
            ex.printStackTrace();
        }
    }
}
```

第二步：更新 DefaultServlet 类，增加异步输出功能。

代码 5-2：DefaultServlet 类

```java
public class DefaultServlet extends HttpServlet {
    //项目内静态资源的根路径
    private final String Static_Resource_PREFIX = "/static";
    //项目外静态资源的根路径
    private final String File_Resource_PREFIX = "D:/ImageOutside";
    @Override
    protected void doGet(HttpServletRequest req,
                         HttpServletResponse resp)
                         throws ServletException, IOException {
        //随机选择阻塞或者非阻塞式输出
        int iChoice = new Random().nextInt(2);
        if(iChoice == 0) {
            System.out.println("阻塞式输出 DefaultServlet ..." +
                                    req.getRequestURI());
            syncOutputResponse(req, resp);
```

```java
        }else{
            System.out.println("非阻塞式输出 DefaultServlet ..." +
                                       req.getRequestURI());
            asyncOutputResponse(req, resp);
        }
    }
    private void asyncOutputResponse(HttpServletRequest req,
                                     HttpServletResponse resp)
                                throws IOException {
        //创建所请求资源的输入流
        InputStream is = getInputStreamFromRequestedResource(req);
        //创建WriteListener监听器,输入流将作为数据源
        WriteListenerForResponse writeListener =
                            new WriteListenerForResponse(is, req, resp);
        //启动监听器
        writeListener.setup();
    }
    private void syncOutputResponse(HttpServletRequest req,
                                    HttpServletResponse resp)
                                throws IOException {
        //获取Response的输出流
        OutputStream os = resp.getOutputStream();
        //获取资源文件的输入流
        InputStream is = getInputStreamFromRequestedResource(req);
        //将输入流的字节序列转入输出流
        is.transferTo(os);
        //关闭输入流
        is.close();
        //冲刷输出流缓冲区
        os.flush();
    }
    //其余省略
}
```

第三步：更新 ThymeleafUtil 类，增加网页的异步输出功能。

代码 5-3：ThymeleafUtil 类

```java
public class ThymeleafUtil {
    //ServletContext存放模板引擎的属性名称
    private static final String TEMPLATE_ENGINE_ATTR =
                "cn.shanghai.cxiao.thymeleaf.TemplateEngineInstance";
    static public void outputHTML(HttpServletRequest req,
                                  HttpServletResponse resp,
                                  Map<String, Object> map,
                                  String templateName)
                                throws IOException {
```

```java
        //随机选择阻塞或者非阻塞式输出
        int iChoice = new Random().nextInt(2);
        if (iChoice == 0) {
            System.out.println("阻塞式输出 HTML ..." + templateName);
            ThymeleafUtil.synchronizedOutputHTML(req, resp, map, templateName);
        }else{
            System.out.println("非阻塞式输出 HTML ..." + templateName);
            ThymeleafUtil.asynchronizedOutputHTML(req, resp, map, templateName);
        }
    }
    static private void asynchronizedOutputHTML(HttpServletRequest req,
                            HttpServletResponse resp,
                            Map<String, Object> map,
                            String templateName)
                            throws IOException{
        //从网页模板生成网页字符串
        String htmlText = ThymeleafUtil.generateHTML(req,resp,map,templateName);
        //创建网页字符串的输入流
        InputStream is = new ByteArrayInputStream(htmlText.getBytes("utf-8"));
        //创建WriteListener监听器，输入流将作为数据源
        WriteListenerForResponse writeListener =
                            new WriteListenerForResponse(is, req, resp);
        //启动监听器
        writeListener.setup();
    }
    static private String generateHTML(HttpServletRequest req,
                            HttpServletResponse resp,
                            Map<String, Object> map,
                            String templateName){
        //从ServletContext的属性取得模板引擎
        ITemplateEngine templateEngine = (ITemplateEngine) req
                .getServletContext()
                .getAttribute(ThymeleafUtil.TEMPLATE_ENGINE_ATTR);
        //使用Thymeleaf构建当前ServletContext的WebExchange
        IWebExchange webExchange = JakartaServletWebApplication
                            .buildApplication(req.getServletContext())
                            .buildExchange(req, resp);
        //创建Thymeleaf的WebContext对象
        WebContext context = new WebContext(webExchange, req.getLocale());
        if(map!=null) {
            for(String key : map.keySet()) {
                //把需要在HTML模板内展示的数据添加至WebContext
                context.setVariable(key, map.get(key));
            }
        }
```

```java
        //使用模板引擎,根据HTML网页模板生成网页字符串
        return templateEngine.process(templateName, context);
    }
    static private void synchronizedOutputHTML(HttpServletRequest req,
                    HttpServletResponse resp,
                    Map<String, Object> map,
                    String templateName)
                    throws IOException {
        //从ServletContext的属性取得模板引擎
        ITemplateEngine templateEngine = (ITemplateEngine) req
                .getServletContext()
                .getAttribute(ThymeleafUtil.TEMPLATE_ENGINE_ATTR);
        //使用Thymeleaf构建当前ServletContext的WebExchange
        IWebExchange webExchange = JakartaServletWebApplication
                .buildApplication(req.getServletContext())
                .buildExchange(req, resp);
        //创建Thymeleaf的WebContext对象
        WebContext context = new WebContext(webExchange, req.getLocale());
        if(map!=null) {
            for(String key : map.keySet()) {
                //把需要在HTML模板内展示的数据添加至WebContext
                context.setVariable(key, map.get(key));
            }
        }
        //使用模板引擎,把HTML模板的处理结果通过Response发送
        templateEngine.process(templateName, context, resp.getWriter());
    }
    //其余省略
}
```

第四步:运行程序,之后在浏览器地址栏中输入 http://localhost:8080/index,查看显示结果。

5.4 用Spring实现

设计思路

Spring 本身并不提供非阻塞输出的设置功能,但提供了非阻塞输出的一种替代机制,即异步应答。采用异步应答时,IndexController 类的方法可以返回一个 Callable<String>对象,该对象代表一个新线程,对浏览器的请求进行真正处理。实现步骤如下:

第一步:修改 IndexController 类。

代码 5-4:IndexController 类

```java
@Controller
public class IndexController {
    @GetMapping("/index")
    public Callable<String> index(){
```

```
            return ()-> "index/main";
    }
}
```

第二步：运行程序，之后在浏览器地址栏中输入 http://localhost:8081/index，查看显示结果。

5.5 用SpringBoot实现

设计思路

SpringBoot 亦是采用异步应答代替非阻塞输出，这一点与 Spring 相同。实现步骤如下：

第一步：修改 IndexController 类，与 5.4 节的第一步相同。

第二步：运行程序，之后在浏览器地址栏中输入 http://localhost:8082/index，查看显示结果。

5.6 小结

本章介绍如何进行 Servlet 应答的非阻塞输出。它是通过为应答输出流设置 WriteListener 来实现的，必须先启动 Request 的异步模式。非阻塞输出可以提高服务器对请求的处理速度。Spring 与 SpringBoot 用异步应答代替 Servlet 的非阻塞输出。

5.7 习题

用户在浏览器地址栏中输入 http://localhost:8080/?n=15&c=马，浏览器显示效果如图 5-4 所示。

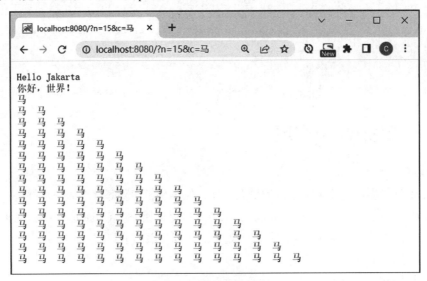

图5-4　浏览器显示效果

说明：地址栏中 n 与 c 是参数名，n 表示三角形的行数，c 表示三角形内的字符。要求采用非阻塞输出。

第 6 章 分派请求

本章介绍如何把请求分派给其他 Servlet 处理。本章案例是通过同步分派（forward）或者异步分派（dispatch）把输出 HTML 网页的工作交由专门的 Servlet 完成。

通过学习本章内容，读者将可以：
- 区分同步委托与异步委托
- 区分转发与分包
- 把应答操作封装到自定义的专门 DisplayServlet

6.1 相关概念

为了提高 Servlet 的复用性，一个具有特定功能的 Servlet 可以接受其他 Servlet 的委托以完成特定的任务。委托者可以等待特定任务的完成，亦可不等待而去处理其他事情；前者通过 RequestDispatcher 对象的 include(.)方法或者 forward(.)方法实现，后者通过 AsyncContext 对象的 dispatch(.)方法实现。

6.1.1 委托的分类

Servlet 可以把请求委托（又称分派）给另一个 Servlet 处理。委托分为同步委托和异步委托，前者的委托者 Servlet 必须等待受委托的 Servlet 处理完毕再继续自己的处理，后者的委托者 Servlet 在移交请求至受委托的 Servlet 后马上返回，继续自己的处理，不会等待受托 Servlet 的处理。委托的分类如图 6-1 所示。

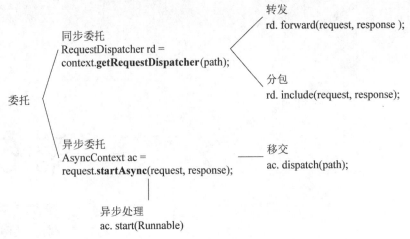

图6-1　委托的分类

同步委托需要创建 RequestDispatcher 接口的实例。RequestDispatcher 接口提供机制支持以下操作：

- 把对请求的处理转发至另一个 Servlet
- 把另一个 Servlet 的输出包含在当前应答中

异步委托需要启动异步环境 AsyncContext，之后可以调用其 dispatch(.)方法将请求移交给另一个 Servlet。

委托操作可以提高 Servlet 复用性。

6.1.2　获取RequestDispatcher对象

通过以下方法可以获得实现 RequestDispatcher 接口的对象：

- ServletContext 的 getRequestDispatcher(String path)
- ServletContext 的 getNamedDispatcher(String servletName)
- ServletRequest 的 getRequestDispatcher(String path)

第一个方法的参数 path 以 ServletContext 容器路径为相对路径，而第三个方法的参数 path 以当前 Servlet 路径为相对路径。

例如，ServletContext 容器路径为 "/"，Servlet 路径模式是 "/spring/*"，假设当前请求路径是 "/spring/cars.html"，则语句 servletRequest.getRequestDispatcher("river.html")的效果等同于语句 servletContext.getRequestDispatcher("/spring/river.html")。

不论是 ServletContext 还是 ServletRequest，它们的 getRequestDispatcher(.)方法允许在路径参数中包括可选的查询字符串。例如以下获取 RequestDispatcher 的代码：

```
String path = "/order?orderno=5";
RequestDispatcher rd = context.getRequestDispatcher(path);
rd.include(request, response);
```

用于创建 RequestDispatcher 的查询字符串参数优先于传递给目标 Servlet 的其他同名参数，与 RequestDispatcher 对象相关的参数只适用于 include(.)方法或者 forward(.)方法调用期间。

6.1.3 使用RequestDispatcher对象

Servlet 调用 RequestDispatcher 接口的 include(.)方法或者 forward(.)方法把请求委托给其他 Servlet 处理。这两个方法的参数或者是委托者 Servlet 的 service(.)方法的传入参数 request 和 response，或者是其包装器对象。Servlet 容器会保证委托给目标 Servlet 的 request 和 response 与当前 Servlet 的 request 和 response 运行于同一个线程。

6.1.4 分派方法的区别

Servlet 可以在任何时候调用 RequestDispatcher 接口的 include(.)方法。include(.)方法的目标 Servlet 可以访问 ServletRequest 对象的所有功能，但是目标 Servlet 只能使用 ServletResponse 对象的 ServletOutputStream 或者 Writer 写出信息，亦可通过显式调用 ServletResponse 的 flushBuffer()方法提交 ServletResponse。目标 Servlet 任何试图设置 ServletResponse 头部信息的操作会被忽略。

Servlet 只有在 ServletResponse 信息还未提交至浏览器时才能调用 RequestDispatcher 接口的 forward(.)方法，此时 forward(.)方法会删除 ServletResponse 缓冲区内未被提交的数据并激活目标 Servlet。在目标 Servlet 中应答信息必须被发送和提交，当目标 Servlet 结束时 ServletResponse 将关闭，除非在目标 Servlet 中由 ServletRequest 开启了异步模式。

若要进行异步分派，则 Servlet 先从其参数 ServletRequest 对象获得 AsyncContext 实例（即开启异步模式），再调用此 AsyncContext 对象的 dispatch(.)方法，此方法会立即返回，接着执行完调用者 Servlet（含过滤器链的反向退出）再去执行目标 Servlet（含其过滤器链），因此调用者 Servlet 与目标 Servlet 均可在任何时候使用 ServletResponse 进行输出，若目标 Servlet 未再开启异步模式则其运行结束时 ServletResponse 将关闭；Servlet 也可在获得 AsyncContext 实例后调用该实例的 start(Runnable)方法，此方法使得另一个异步的线程开始执行 Runnable 过程，在 Runnable 中调用 AsyncContext 对象的 complete()方法才能关闭 ServletResponse。另外，如果 start(Runnable)与 dispatch(.)组合使用，即在 Runnable 过程中调用 dispatch(.)，则 Runnable 以一个新线程运行，而目标 Servlet 将以另一个新线程运行，此时无须在 Runnable 中调用 AsyncContext 的 complete()方法，因为目标 Serlvet 运行完毕将自动关闭 ServletResponse。

分派方法汇总见表 6-1。

表 6-1 分派函数汇总表

方法	.include(.)	.forward(.)	.dispatch(.)	.sendError(.)	.sendRedirect(.)

调用代码	getServletContext().getRequestDispatcher("/xyz?abc=Fudan").include(req, resp);
	getServletContext().getRequestDispatcher("/xyz?abc=Fudan").forward(req, resp);
	req.startAsyncContext(req, resp).dispatch("/xyz?abc=Fudan");
	resp.sendError(500, "message"); // 已配置 new ErrorPage().setLocation("/xyz?abc=Fudan");
	resp.sendRedirect("/xyz?abc=Fudan");				

同/异步	同步	同步	异步	异步	异步
在调用方法之后，调用者的Response是否关闭	不关闭	关闭	不关闭	关闭	关闭
在调用方法之前、之后（包括调用者的Filter中）从Response输出	输出有效，且可在调用前flush	输出无效，若调用前flush，则forward(.)报错	输出有效，且可在调用前flush	输出无效，若调用前flush，则sendError(.)报错	输出无效，若调用前flush，则sendRedirect(.)报错
Target Servlet（含其Filter）执行完毕Response是否关闭	Response 不关闭	Response 关闭	Response 关闭（注：先执行完调用者（含Filter链的反向退出）再执行Target Servlet（及其Filter），所以可以在调用前后写入Response）	Response 关闭	Response 关闭
在Target Servlet中调用 request.getRequestURI() 是否正是其URI	不是/xyz，而是req.getRequestURI()	正是 /xyz	正是 /xyz	正是 /xyz	正是 /xyz
在Target Servlet中调用 request.getParameter("abc")	Fudan	Fudan	Fudan	Fudan	Fudan

6.2 案例描述

在浏览器地址栏中输入网址 http://localhost:8080/index，浏览器显示效果如图 6-2 所示。

图6-2 案例运行界面

说明：在此案例中使用转发与移交。

6.3 用Servlet实现

设计思路

构建专门的DisplayServlet，用于接受对页面进行显示的功能委托。委托者通过ServletRequest的命名属性向DisplayServlet传递参数。DisplayServlet可以接受异步委托（dispatch），也可以接受同步委托（forward），委托方式由ThymeleafUtil 的 outputHTML(.)方法随机选择。委托的函数调用关系如图 6-3 所示。DisplayServlet 接受委托后再随机选择阻塞式或非阻塞式输出完成应答。实现步骤如下：

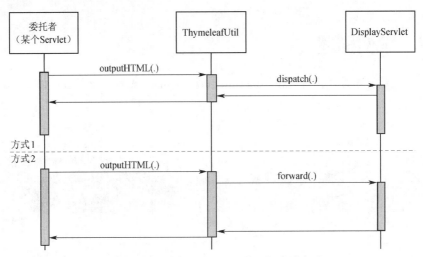

图6-3 DisplayServlet接受委托的两种方式

第一步：新增 DisplayServlet 类。

代码 6-1：DisplayServlet 类

```java
public class DisplayServlet extends HttpServlet {
    public static final String EACH_RESP_THYMELEAF_MAP_DATA_ATTR =
                "cn.shanghai.cxiao.response.Thymeleaf_data";
    public static final String EACH_RESP_TEMPLATE_NAME_ATTR =
                "cn.shanghai.cxiao.template.name";
    public static final String DISPLAY_SERVLET_URI = "/display";
    public static final String DISPLAY_SERVLET_NAME = "display.servlet";
    @Override
    protected void doGet(HttpServletRequest req, HttpServletResponse resp)
    throws ServletException, IOException {
        //通过request Attribute 获取传入参数
        Map<String, Object> map = (Map<String, Object>)req.getAttribute(
                DisplayServlet.EACH_RESP_THYMELEAF_MAP_DATA_ATTR);
        String templateName = (String)req.getAttribute(
                DisplayServlet.EACH_RESP_TEMPLATE_NAME_ATTR);
        //随机选择阻塞或者非阻塞式输出
        int iChoice = new Random().nextInt(2);
        if(iChoice == 0) {
            System.out.println("\t阻塞式输出 HTML ..." + templateName);
            ThymeleafUtil.synchronizedOutputHTML(req, resp, map, templateName);
        }else{
            System.out.println("\t非阻塞式输出 HTML ..." + templateName);
            ThymeleafUtil.asynchronizedOutputHTML(req, resp, map, templateName);
        }
    }
    @Override
    protected void doPost(HttpServletRequest req, HttpServletResponse resp)
```

```
        throws ServletException, IOException {
            doGet(req, resp);
        }
    }
```

第二步：在 ServletInitializer 中向 Servlet 容器注册 DisplayServlet。

代码 6-2：注册 DisplayServlet

```
//注册 DisplayServlet
servlet = new DisplayServlet();
registration = servletContext.addServlet(
                    DisplayServlet.DISPLAY_SERVLET_NAME, servlet);
registration.setLoadOnStartup(4);
registration.addMapping(DisplayServlet.DISPLAY_SERVLET_URI);
registration.setAsyncSupported(true);
```

第三步：更新 ThymeleafUtil 的 outputHTML(.)方法。

代码 6-3：outputHTML(.)方法

```
public class ThymeleafUtil {
    //ServletContext 存放模板引擎的属性名称
    private static final String TEMPLATE_ENGINE_ATTR =
                "cn.shanghai.cxiao.thymeleaf.TemplateEngineInstance";

    static public void outputHTML(HttpServletRequest req,
                                  HttpServletResponse resp,
                                  Map<String, Object> map,
                                  String templateName)
                                  throws IOException, ServletException {
        req.setAttribute(DisplayServlet.EACH_RESP_THYMELEAF_MAP_DATA_ATTR, map);
        req.setAttribute(DisplayServlet.EACH_RESP_TEMPLATE_NAME_ATTR,templateName);
        //委托时有两种选择，故采用随机方式
        int iChoice = new Random().nextInt(2);
        if (iChoice==0) {
            //异步 dispatch
            System.out.println("Async dispatch..." + req.getRequestURI());
            AsyncContext asct = req.startAsync(req, resp);
            asct.dispatch(DisplayServlet.DISPLAY_SERVLET_URI);
        }else{
            //同步 forward
            System.out.println("Sync forward..." + req.getRequestURI());
            req.getServletContext()
                .getRequestDispatcher(DisplayServlet.DISPLAY_SERVLET_URI)
                .forward(req, resp);
        }
    }
//其余省略
}
```

第四步：运行程序，之后在浏览器地址栏中输入http://localhost:8080/index，查看显示结果。

6.4 用Spring实现

设计思路

使用Spring时如果把委托者Controller的方法参数定义为HttpServletRequest和HttpServletResponse（此时无须返回值），那么可以用RequestDispatcher或者AsyncContext分派请求。如果不用这种方式，那么可以采用Spring所封装的forward，也可以采用Spring所封装的redirect。因此这里提供四种方法。先定义一个只用于显示网页的DisplayController，委托者将处理的结果数据保存于Model或其子类对象，再转发或重定向至DisplayController。实现步骤如下：

第一步：新增DisplayController类。

代码6-4：DisplayController类

```
@Controller
public class DisplayController {
    @RequestMapping("/display")
    public String abc(@RequestParam String templatename, Model model){
        //参数templatename指明网页模板名称，由委托者传入
        //参数model用于存储要显示的数据，由委托者传入
        //此处还可往model中添加其他数据
        //这里只需要返回网页模板名称就可以将model中的数据显示
        return templatename;
    }
}
```

第二步：修改IndexController类。

代码6-5：IndexController类

```
@Controller
public class IndexController {
    @GetMapping("/index")           //方法1
    public String index(RedirectAttributes model){
        //将要显示的数据存入RedirectAttributes
        //model会传递至重定向的目标Controller
        model.addFlashAttribute("name", "HelloWorld from index()!");
        //把网页模板名作为参数重定向给DisplayController进行显示处理
        return "redirect:/display?templatename=index/main";
    }
    @GetMapping("/index1")          //方法2
    public String index1(Model model){
        //model数据会被转发至目标Controller
        model.addAttribute("name", "HelloWorld from index1()! ");
        //把网页模板名作为参数转发给DisplayController进行显示处理
```

```java
        return "forward:/display?templatename=index/main";
    }
    @GetMapping("/index2")              //方法 3
    public void index2(HttpServletRequest req, HttpServletResponse resp)
    throws ServletException, IOException {
        //目标 Controller 的 model 数据来自 req 的命名属性
        req.setAttribute("name", "HelloWorld from index2()! ");
        RequestDispatcher requestDispatcher = req.getServletContext()
                .getRequestDispatcher("/display?templatename=index/main");
        //把网页模板名作为参数转发给 DisplayController 进行显示处理
        requestDispatcher.forward(req, resp);
    }
    @GetMapping("/index3")              //方法 4
    public void index3(HttpServletRequest req, HttpServletResponse resp){
        //目标 Controller 的 model 数据来自 req 的命名属性
        req.setAttribute("name", "Jakarta from index3()! ");
        AsyncContext asyncContext = req.startAsync(req, resp);
        //把网页模板名作为参数移交给 DisplayController 进行显示处理
        asyncContext.dispatch("/display?templatename=index/main");
    }
}
```

index(.)方法的参数类型是 RedirectAttributes 而不是常规的 Model，因为只有这样才能把数据传递至重定向的目标 Controller。

第三步：修改 resources\templates\index\main.html 文件。

```html
//第一处修改:
<html xmlns:th="http://www.thymeleaf.org">
//第二处修改:
<p class="m-0" th:text="${name}" />
```

第二处修改的${name}正是第二步调用 addFlashAttribute(.)方法所用的属性名称。

第四步：运行程序，之后在浏览器地址栏中输入 http://localhost:8081/index，查看显示结果；再把地址栏中的 index 分别改成 index1、index2、index3，查看显示结果。

6.5　用SpringBoot实现

设计思路

与 Spring 实现方法相同。实现步骤如下：

第一步：新增 DisplayController 类，与 6.4 节第一步相同。

第二步：修改 IndexController 类，与 6.4 节第二步相同。

第三步：修改 resources\templates\index\main.html 文件，与 6.4 节第三步相同。

第四步：运行程序，之后在浏览器地址栏中输入 http://localhost:8082/index，查看显示结果；再把地址栏中的 index 分别改成 index1、index2、index3，查看显示结果。

6.6 小结

HTML 网页的显示功能可以同步或者异步委托给专门的自定义 DisplayServlet 来完成，由 DisplayServlet 随机选择阻塞式或非阻塞式输出应答。委托者通过 ServletRequest 的命名属性把数据传递给 DisplayServlet。而 Spring、SpringBoot 则可以把请求分派给其他 Controller，Controller 之间通过 Model 传递数据。

6.7 习题

修改本章案例，使得在浏览器地址栏中输入 http://localhost:8080/index 时，显示效果如图6-4所示，要求采用 include(.)方法实现。

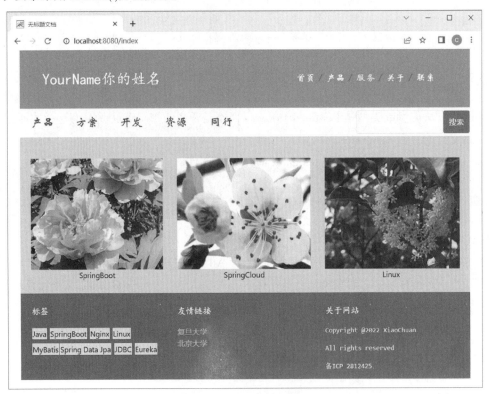

图6-4 浏览器显示效果

说明：三个素材文件位于"源码\chapter06\习题附件"文件夹内。

第 7 章

会 话

本章介绍会话类 HttpSession。本章案例是实现具有会话特性的多次访问。
通过学习本章内容，读者将可以：
- 添加过滤器以消除在 URL 中自动生成的 jsessionid
- 使用 HttpSession 管理会话

7.1 相关概念

HTTP 是无状态协议，浏览器连续两次请求之间并没有状态依赖，但是把来自某个浏览器的多个连续请求进行逻辑上的关联（即构成一次会话）对于构建有效的 Web 应用是很重要的。HttpSession 接口允许 Servlet 容器使用该接口的方法来跟踪用户会话。

7.1.1 会话跟踪机制

常见的会话跟踪机制有 Cookie 和 URL 重写。通过 HTTP Cookie 进行会话跟踪是最常用的会话跟踪机制，所有 Servlet 容器都支持 Cookie。

一次会话中，Servlet 容器在答复浏览器的第一次请求时向浏览器发送一个 Cookie，浏览器在随后的每次请求中都包含此 Cookie，因此这些请求通过共享此 Cookie 形成一次会话。这个用于会话跟踪的 Cookie 的标准名称是 JSESSIONID。

URL 重写是会话跟踪的最低标准。当浏览器不接受 Cookie 时，服务端会在应答页面发送之前为页面内的每个内部 URL 添加一个名为 jsessionid 的参数，参数值就是服务端为这次会话创建的 Session ID。以下是把 Session ID 编码成 URL 中 jsessionid 参数的一个例子：

```
http://www.spring.com.cn/machine/index.html;jsessionid=83536
```

在 URL 中包含 Session ID 不是一个明智之举，因为这将为攻击者提供便利。

7.1.2 HttpSession

HttpSession 对象在 ServletContext 范围内有效,同一个 ServletContext 内的不同 Servlet 之间可以共享 HttpSession 对象,而不同 ServletContext 之间不可共享 HttpSession 对象。Servlet 可以通过自定义的名称把某个对象关联至 HttpSession 对象,使该对象成为 HttpSession 的命名属性。绑定至 HttpSession 的任何对象对于同一个 ServletContext 的其他 Servlet 都是可用的。不同的 Servlet 所处理的具有相同 HttpSession ID 的多个请求在逻辑上组成一个会话。

当把实现了 HttpSessionBindingListener 接口的对象本身绑定至 HttpSession 或者从 HttpSession 解除此对象绑定时,将会触发此对象中 HttpSessionBindingListener 接口的相应方法。与之不同的是,只要有对象绑定至 HttpSession 或者从 HttpSession 解除对象绑定又或者替换 HttpSession 的绑定对象,就会触发 HttpSessionAttributeListener 监听器的相应方法,不论该对象是否实现了 HttpSessionAttributeListener 接口。

在 HTTP 协议中,当浏览器不再活动时,不会发送显式的终结信号。这意味着能向服务端表明客户端不再活动的唯一机制只能是超时期限。

HttpSession 对象的默认超时期限是由 Servlet 容器定义的,可以通过调用 ServletContext 接口的 getSessionTimeout()方法或者 HttpSession 接口的 getMaxInactiveInterval()方法获取超时期限。另外,也可以通过调用 ServletContext 接口的 setSessionTimeout(minutes)方法或者 HttpSession 接口的 setMaxInactiveInterval(seconds)方法重新设置超时期限。如果超时期限设置为 0 或者负数,那么 HttpSession 对象将永不过期。当调用 HttpSession 接口的 invalidate()方法时,只有在使用了此 HttpSession 的所有 Servlet 都退出其 service(.)方法之后,HttSession 对象才会失效。一旦 HttpSession 对象失效,则新的请求就看不到这个 HttpSession。

ServletContext 接口会保证代表 Session Attributes 的内部数据结构以线程安全的方式操作,而开发者负责以线程安全的方式访问 Attribute 对象本身。

7.1.3 在Thymeleaf的URL中传递参数

表 7-1 列出了在 Thymeleaf 的 URL 中传递参数的方法。

表 7-1 在 URL 中传递参数的方法

	查询参数(QueryString)
单参数	`<a th:href="@{/abc(id=${user.userId})}">跳转</a>`
多参数	`<a th:href="@{/abc(id=${user.userId}, name=${user.name})}">跳转</a>`
	路径参数(Path Parameter)
单参数	`<a th:href="@{/abc/{id}(id=${user.userId})}">跳转</a>`
多参数	`<a th:href="@{/abc/{id}/{name}(id=${user.userId}, name=${user.name})}">跳转</a>`

7.2 案例描述

在浏览器地址栏中输入网址 http://localhost:8080/index,浏览器显示效果如图 7-1 所示。

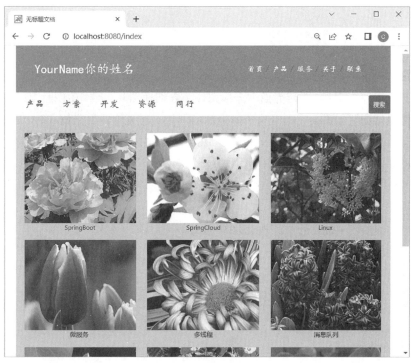

图7-1 案例运行界面（1）

说明：页面上每张图片是一个超链接，当单击某张图片时，该图片会切换成另一张图片，图 7-2 所示为切换三张图片之后的页面效果。

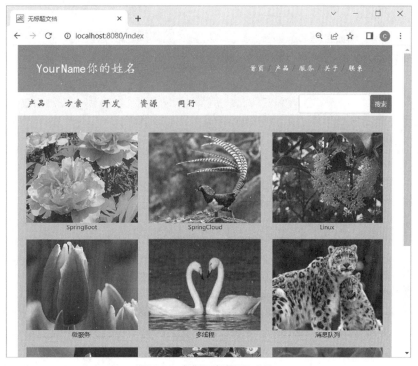

图7-2 案例运行界面（2）

7.3 用Servlet实现

设计思路

在 IndexServlet 中存储浏览器界面上全部图片（含切换图片）的 URI 信息，在 HttpSession 中以命名属性存储浏览器界面上每个格子所显示的图片。当浏览器发出请求时，IndexServlet 根据请求时的参数更改所点击的界面格子对应于 HttpSession 中所记录的图片，将其替换成要切换的新图片，最后把 HttpSession 中全部图片的 URI 信息发送至浏览器，由浏览器展示图片。为了降低内存的使用，HttpSession 中并不存放图片本身，而是存放用来标识图片的序号。实现步骤如下：

第一步：在模块的 pom.xml 中添加 lombok 依赖。

代码 7-1：lombok 依赖

```xml
<dependency>
    <groupId>org.projectlombok</groupId>
    <artifactId>lombok</artifactId>
    <version>1.18.24</version>
</dependency>
```

第二步：添加辅助类 ImageInfo。

代码 7-2：ImageInfo 类

```java
@Data
@AllArgsConstructor
@NoArgsConstructor
class ImageInfo{
    private String imageUri;
    private String label;
}
```

该类用于存储在 HTML 模板页面中每个格子要显示的信息。

第三步：修改已有的 IndexServlet 类。

代码 7-3：IndexServlet

```java
public class IndexServlet extends HttpServlet {
    //以 9 行 3 列的结构存储全部图片及其对应的切换图片，第 3 列是文字标签
    //每一行对应界面上的一格
    private final List<List<String>> IMAGES = List.of(
        //项目文件夹 resources\static\image\portfolio\内已有 port01~09.jpg
        //文件系统 D:\ImageOutside\main\内已有 s2~s10.PNG
        //这些图片文件已经映射成如下 URI
        List.of("/image/portfolio/port01.jpg", "/pic/main/s3.PNG","SpringBoot"),
        List.of("/image/portfolio/port02.jpg", "/pic/main/s10.PNG","SpringCloud"),
        List.of("/image/portfolio/port03.jpg", "/pic/main/s9.PNG","Linux"),
        List.of("/image/portfolio/port04.jpg", "/pic/main/s6.PNG","微服务"),
```

```
        List.of("/image/portfolio/port05.jpg", "/pic/main/s8.PNG", "多线程"),
        List.of("/image/portfolio/port06.jpg", "/pic/main/s7.PNG", "消息队列"),
        List.of("/image/portfolio/port07.jpg", "/pic/main/s5.PNG", "Python"),
        List.of("/image/portfolio/port08.jpg", "/pic/main/s2.PNG", "机器学习"),
        List.of("/image/portfolio/port09.jpg", "/pic/main/s4.PNG", "Redis")
    );
    @Override
    protected void doGet(HttpServletRequest req, HttpServletResponse resp)
    throws ServletException, IOException {
        int[] imageIdList = (int[]) req.getSession().getAttribute("imageIds");
        if (imageIdList == null){
            //此时imageIdList的每个元素等于0,对应于IMAGES的第1列
            imageIdList = new int[IMAGES.size()];
            //存储于命名属性imageIds,以供下一次请求时使用
            req.getSession().setAttribute("imageIds", imageIdList);
        }
        String path = req.getPathInfo();
        //根据uri的路径参数修改session数据
        if (path!=null) {              //用户单击了界面上某张图片
            //从路径信息解析出格子编号
            int imageId = Integer.valueOf(
                        path.substring(path.lastIndexOf('/') + 1));
            //把此格图片替换成要切换的图片
            imageIdList[imageId] = (imageIdList[imageId] + 1) % 2;
        }
        //根据session数据生成将在页面中显示的数据
        List<ImageInfo> imageInfoList = new ArrayList<>();
        for (int i=0; i<imageIdList.length; i++){
            //从IMAGES获取每个格子要显示的图片的URI及文字标签
            imageInfoList.add(
                new ImageInfo(IMAGES.get(i).get(imageIdList[i]),
                              IMAGES.get(i).get(2)));
        }
        //在浏览器中显示main.html,用map传递要显示的对象信息
        Map<String, Object> map = Map.of("images", imageInfoList);
        String templateName = "index/main";
        ThymeleafUtil.outputHTML(req, resp, map, templateName);
    }
}
```

第四步：更新文件 resources\templates\index\main.html 中显示图片的 Div。

代码 7-4：HTML 文件中显示图片的 Div

```
<div class="col-4" th:each="imageInfo,iterStat:${images}">
  <a th:href="@{/index/{pid}(pid=${iterStat.index})}">
```

```html
    <img th:src="@{${imageInfo.imageUri}}" alt=""
         width="200" height="150"/>
</a>
<p>
  <a th:href="@{/index/{pid}(pid=${iterStat.index})}"
     th:text="${imageInfo.label}" />
</p>
</div>
```

这个 Div 用于批量显示图片。

第五步：在 ServletInitializer 中修改 IndexServlet 的映射 URI。

代码 7-5：IndexServlet 的配置

```
//注册 IndexServlet 并作配置
servlet = new IndexServlet();
registration = servletContext.addServlet("index.servlet", servlet);
registration.setLoadOnStartup(3);
registration.addMapping("/index/*");
registration.setAsyncSupported(true);
```

把 addMapping(.)方法的参数由之前的"/index"更改为"/index/*"，以便处理"/index/3"之类 URI 请求。

第六步：创建 UrlJSessionIdFilter 类，用于消除服务端应答时自动在 URL 上附加的 jsessionid 参数。

代码 7-6：UrlJSessionIdFilter 类

```java
public class UrlJSessionIdFilter implements Filter {
    @Override
    public void doFilter(ServletRequest servletRequest,
                         ServletResponse servletResponse,
                         FilterChain filterChain)
                    throws IOException, ServletException {
        if(!(servletRequest instanceof HttpServletRequest)) {
            filterChain.doFilter(servletRequest, servletResponse);
            return;
        }
        HttpServletResponse httpResponse =
                        (HttpServletResponse) servletResponse;
        HttpServletResponseWrapper wrappedResponse =
              new HttpServletResponseWrapper(httpResponse){
            //重写原先的两个方法，把其中的 encodeURL(.)调用去掉，直接返回 url
            @Override
            public String encodeRedirectURL(String url) {
                return url;
            }
```

```
            @Override
            public String encodeURL(String url) {
                return url;
            }
        };
        filterChain.doFilter(servletRequest, wrappedResponse);
    }
}
```

当浏览器第一次请求时，Servlet 容器创建一个新的 HttpSession 对象，Servlet 容器并不确定浏览器是否支持 Cookie，所以它重写应答页面内的 URL，为其自动添加了一个 jsessionid 参数，其值就是 Servlet 容器产生的 HttpSession 的 ID 值。

当浏览器第二次请求时如果带着 Cookie，服务器就知道重写 URL 不是必需的，所以不会继续在 URL 中添加 jsessionid 参数；如果浏览器第二次请求时没有带着 Cookie，服务器就会继续在 URL 中添加 jsessionid 参数。

第七步：在 ServletInitializer 中注册 UrlJSessionIdFilter。

代码 7-7：注册 UrlJSessionIdFilter

```
//注册 UrlJSessionIdFilter 过滤器
filter = new UrlJSessionIdFilter();
filterRegistration = servletContext.addFilter("sessionFilter", filter);
filterRegistration.addMappingForUrlPatterns(null, true, "/*");
```

第八步：运行程序，之后在浏览器地址栏中输入 http://localhost:8080/index，查看显示结果。

7.4 用Spring实现

设计思路

在 IndexController 中存储浏览器界面上全部图片（含切换图片）的 URI 信息，在 HttpSession 中以命名属性存储浏览器界面上每个格子所显示的图片。当浏览器发出请求时，IndexController 根据请求时的参数更改所点击的界面格子对应于 HttpSession 中所记录的图片，将其替换成要切换的新图片，最后把 HttpSession 中全部图片的 URI 信息发送至浏览器，由浏览器展示图片。为了降低内存的使用，HttpSession 中并不存放图片本身，而是存放用来标识图片的序号。实现步骤如下：

第一步：在模块的 pom.xml 中添加 lombok 依赖。

代码 7-8：lombok 依赖

```
<dependency>
    <groupId>org.projectlombok</groupId>
    <artifactId>lombok</artifactId>
    <version>1.18.24</version>
</dependency>
```

第二步：添加辅助类 ImageInfo。

代码 7-9：ImageInfo 类

```
@Data
@AllArgsConstructor
@NoArgsConstructor
class ImageInfo{
    private String imageUri;
    private String label;
}
```

该类用于存储在 HTML 模板页面中每个格子要显示的信息。

第三步：修改已有的 IndexController 类。

代码 7-10：IndexController 类

```
@Controller
public class IndexController {
    private final List<List<String>> IMAGES = List.of(
        //项目文件夹 resources\static\image\portfolio\内已有 port01~09.jpg
        //文件系统 D:\ImageOutside\main\内已有 s2~s10.PNG
        //这些图片文件已经映射成如下 URI
        List.of("/image/portfolio/port01.jpg", "/pic/main/s3.PNG","SpringBoot"),
        List.of("/image/portfolio/port02.jpg", "/pic/main/s10.PNG","SpringCloud"),
        List.of("/image/portfolio/port03.jpg", "/pic/main/s9.PNG","Linux"),
        List.of("/image/portfolio/port04.jpg", "/pic/main/s6.PNG","微服务"),
        List.of("/image/portfolio/port05.jpg", "/pic/main/s8.PNG","多线程"),
        List.of("/image/portfolio/port06.jpg", "/pic/main/s7.PNG","消息队列"),
        List.of("/image/portfolio/port07.jpg", "/pic/main/s5.PNG","Python"),
        List.of("/image/portfolio/port08.jpg", "/pic/main/s2.PNG","机器学习"),
        List.of("/image/portfolio/port09.jpg", "/pic/main/s4.PNG","Redis")
    );
    @GetMapping("/index")
    public String index(Model model, HttpSession session){
        return index(null, model, session);
    }
    @GetMapping("/index/{id}")
    public String index(@PathVariable("id") Integer imageId,
                        Model model,
                        HttpSession session){
        int[] imageIdList = (int[])session.getAttribute("imageIds");
        if (imageIdList == null){
            imageIdList = new int[IMAGES.size()];
            session.setAttribute("imageIds", imageIdList);
        }
        //根据 uri 的路径参数修改 session 数据
```

```
        if (imageId!=null) {
            imageIdList[imageId] = (imageIdList[imageId] + 1) % 2;
        }
        //根据session数据生成将在页面中显示的数据
        List<ImageInfo> imageInfoList = new ArrayList<>();
        for (int i=0; i<imageIdList.length; i++){
            imageInfoList.add(
                new ImageInfo(IMAGES.get(i).get(imageIdList[i]),
                              IMAGES.get(i).get(2)));
        }
        model.addAttribute("images", imageInfoList);
        return "index/main";
    }
}
```

第四步：更新文件 resources\templates\index\main.html 中显示图片的 Div。

代码 7-11：HTML 文件中显示图片的 Div

```
<div class="col-4" th:each="imageInfo,iterStat:${images}">
  <a th:href="@{/index/{pid}(pid=${iterStat.index})}">
    <img th:src="@{${imageInfo.imageUri}}"
         alt="" width="200" height="150"/>
  </a>
  <p>
    <a th:href="@{/index/{pid}(pid=${iterStat.index})}"
       th:text="$ {imageInfo.label}" />
  </p>
</div>
```

这个 Div 用于批量显示图片。

第五步：运行程序，之后在浏览器地址栏中输入 http://localhost:8081/index，查看显示结果。

7.5 用SpringBoot实现

设计思路

与 Spring 实现方案的设计思路相同。实现步骤如下：

第一步：在模块的 pom.xml 中添加 lombok 依赖。

代码 7-12：lombok 依赖

```
<dependency>
    <groupId>org.projectlombok</groupId>
    <artifactId>lombok</artifactId>
    <version>1.18.24</version>
</dependency>
```

第二步：添加辅助类 ImageInfo，与 7.4 节第二步相同。

第三步：修改已有的 IndexController 类，与 7.4 节第三步相同。

第四步：更新文件 resources\templates\index\main.html 中批量显示图片的 Div，与 7.4 节第四步相同。

第五步：运行程序，之后在浏览器地址栏中输入 http://localhost:8082/index，查看显示结果。

7.6 小结

HttpSession 把多次连续的请求从逻辑上构成一个整体，称为一次会话。比如从登录系统到退出系统之间的多次请求及其应答就构成一次会话。HttpSession 为一次会话的全部请求提供了时间跨度上的信息共享。会话的持续时间可以由预设的超时期限决定，也可以用代码 HttpSession.invalidate() 显式地使之结束。

7.7 习题

修改本章案例的 IndexServlet 及网页，使得在地址栏中输入 http://localhost:8080/index 时浏览器显示效果如图 7-3 所示。当用户单击"上一张"或者"下一张"文字链接时，中间的图片依次轮换。

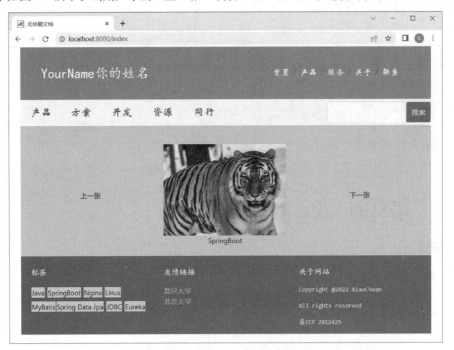

图7-3　浏览器显示效果

提示：用 HttpSession 存储当前图片的序号。

第 8 章

提交表单

本章介绍数据绑定,包括输入数据绑定和输出数据绑定。输入数据绑定是把表单提交的参数转换成对象属性,输出数据绑定是把对象属性回显至表单元素或者页面元素。本章案例是实现用户信息录入及修改。

通过学习本章内容,读者将可以:
- ❏ 把表单元素转换成对象的属性
- ❏ 用 Thymeleaf 显示对象属性

8.1 相关概念

当提交表单时,表单数据以文本格式通过 HTTP 的 POST 方法被提交至服务器,而 Servlet 适宜处理内存中的对象,因此服务端需要将文本数据转换成对象;在 Servlet 进行业务处理时,它所产生的对象需要以文本格式作为 HTTP 应答发送至浏览器,因此服务端又需要将对象转换成文本数据。

8.1.1 请求时字符编码

浏览器在 HTTP 请求的消息头 Content-Type 中指明本次请求的消息体所采用的字符编码,服务端可以通过如下两个方法获取此字符编码设置:
- ❏ ServletRequest 的 getCharacterEncoding()
- ❏ ServletContext 的 getRequestCharacterEncoding()

Servlet 容器根据此字符编码来创建请求消息体的 Reader 以及对 POST 数据进行解码。如果浏览器没有指明所用字符编码并且请求消息体采用的不是默认的 ISO-8859-1 编码,则服务端可以通过如下方法设置在解码请求消息体时所用的字符编码:

- ServletRequest 的 setCharacterEncoding(String enc)
- ServletContext 的 setRequestCharacterEncoding(String enc)

上述两个 set 方法必须在解析 POST 数据之前或者从请求消息体读取任何数据之前调用，而读取任一数据之后再调用上述 set 方法将无效。

8.1.2 输入数据绑定

满足以下条件时表单数据才会提取至 ServletRequest 对象的参数集内：

- 请求是 HTTP 或者 HTTPS 请求
- HTTP 方法是 POST
- 表单提交时 Content-Type 是 application/x-www-form-urlencoded
- Servlet 优先调用 ServletRequest 对象的任何一个访问请求参数的方法，如 getParameter(.)

如果上述条件不满足，表单数据就不会包含于参数集，即无法通过任何一个访问请求参数的方法获取参数值，必须通过 ServletRequest 对象的输入流获取表单数据。如果上述条件满足，将不能从 ServletReqeust 对象的输入流直接读取表单数据。图 8-1 所示为表单数据处理流程的示例。

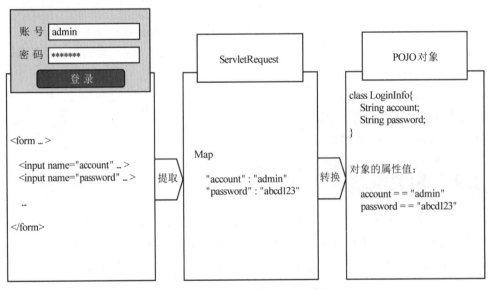

图 8-1 表单数据处理流程示例

8.1.3 输出数据绑定

Thymeleaf 在模板网页中使用 OGNL 表达式显示对象信息。有关 OGNL 语法和功能的详细介绍请自行百度搜索"OGNL 语言指南"。

显示对象信息的变量表达式看起来像这样：

```
<span th:text="${user.sex.desc}" />
```

上面的表达式在 OGNL 中等价于：

```
((User)context.getVariable("user")).getSex().getDesc()
```

这些变量表达式不仅用于输出，还包括更复杂的处理，如条件判断、迭代等。例如：

```
<li th:each="user : ${users}">
```

这里${users}从上下文选择名为 users 的变量，并将其转换为可在 th:each 循环中使用的迭代器。

8.1.4 sendRedirect(.)方法

ServletResponse 对象的 sendRedirect(.)方法是 Servlet 容器对当前请求的一种应答，此方法将设置相应的应答消息头和消息体指示客户端重新向一个不同的 URL 发出请求。sendRedirect(.)方法执行后会提交并终结应答。执行此方法后不应该再向客户端输出数据，因为这些数据将被忽略。

Servlet 可以用相对 URI 路径调用 sendRedirect(.)方法，因为 Servlet 容器会把相对路径转换成绝对路径传回客户端；如果路径转换失败，此方法会抛出 IllegalArgumentException 异常。

如果在调用 sendRedirect(.)方法之前写入 ServletResponse 缓冲区的数据尚未被发送至客户端，此方法将删除缓冲区内已有的数据，并写入方法所设置的数据；如果在调用此方法之前应答已经被提交，此方法会抛出 IllegalStateException 异常。

8.2 案例描述

用户在浏览器地址栏中输入地址 http://localhost:8080/user，浏览器显示效果如图 8-2 所示。

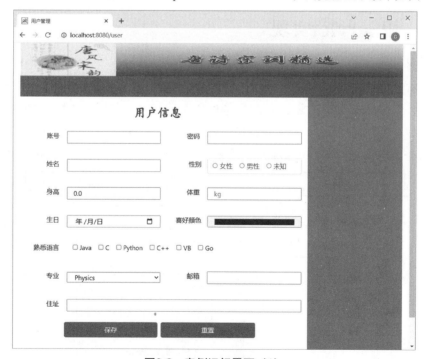

图8-2 案例运行界面（1）

用户输入各项，单击"保存"按钮，之后界面右侧会列出已经保存的用户信息。单击右侧某项，则界面会显示此项的详细信息，效果如图 8-3 所示。

图8-3 案例运行界面（2）

8.3 用Servlet实现

设计思路

根据界面上"性别"、"专业"、"熟悉语言"分别定义枚举类 Gender、Major、PLanguage；定义通用枚举转换器 EnumConversion，提供字符串与枚举值相互转换功能；再基于 EnumConversion 分别对每个枚举类实现转换器的 IConvertor 接口，得到 GenderConvertor 类、MajorConvertor 类、PLanguageConvertor 类。枚举类与转换器的继承关系如图 8-4 所示。

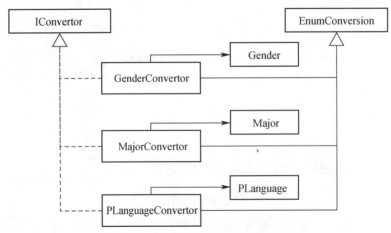

图8-4 几个枚举类与转换器的继承关系

新增自定义注解@FormField，用于标注 User 类中与表单输入域对应的成员变量；在 @FormField 中除了指明表单输入域名称，还能指明所用转换器类型，如 GenderConvertor.class。

定义工具类 ConvertorFactory，提供 mapToObject(.)函数，将"键-值"对的 Map 数据转换成对象的属性值；为了便于转换，ConvertorFactory 中也提供若干把字符串转成常用类型（如 Integer、LocalDate、Color 等）对象的转换函数。

表单数据转换至 POJO 对象的工作在 UserServletRequestListener 中进行。

实现步骤如下：

第一步：新增自定义转换器接口 IConvertor。

代码 8-1：IConvertor 接口

```
package cxiao.sh.cn.convertor;
public interface IConvertor <T>{
    public T convert(String sValue);
}
```

第二步：新增 FormField 注解定义，用于之后标注与表单域对应的属性。

代码 8-2：FormField 注解类

```
@Documented
@Target({FIELD})
@Retention(RUNTIME)
public @interface FormField {
    String name() default "";
    Class<? extends IConvertor> convertorClass() default IConvertor.class;
    String[] convertParams() default { };
}
```

第三步：新增若干枚举类。

代码 8-3：Gender 类

```
public enum Gender {
    FEMALE(0, "Female", "女性"),
    MALE(1, "Male", "男性") ,
    UNSPECIFIED(2, "Unknown", "未知");
    private int index;
    private String enDesc;
    private String cnDesc;
    Gender(int index, String enDesc, String cnDesc){
        this.index = index;
        this.enDesc = enDesc;
        this.cnDesc = cnDesc;
    }
    public int getIndex(){
        return index;
    }
```

```java
    public String getDesc(Locale locale){
        if (locale.equals(Locale.CHINA)){
            return getCnDesc();
        }
        return getEnDesc();
    }
    public String getEnDesc(){
        return enDesc;
    }
    public String getCnDesc(){
        return cnDesc;
    }
}
```

代码 8-4：Major 类

```java
public enum Major {
    PHYSICS(10, "Physics", "物理"),
    CHEMISTRY(11, "Chemistry", "化学") ,
    MATHEMATICS(12, "Mathematics", "数学"),
    ECONOMICS(13, "Economics", "经济学"),
    COMPUTER(14, "Computer", "计算机");
    private int index;
    private String enDesc;
    private String cnDesc;
    Major(int index, String enDesc, String cnDesc){
        this.index = index;
        this.enDesc = enDesc;
        this.cnDesc = cnDesc;
    }
    public int getIndex(){
        return index;
    }
    public String getDesc(Locale locale){
        if (locale.equals(Locale.CHINA)){
            return getCnDesc();
        }
        return getEnDesc();
    }
    public String getEnDesc(){
        return  enDesc;
    }
    public String getCnDesc(){
        return cnDesc;
    }
}
```

代码 8-5：PLanguage 类

```java
public enum PLanguage {
    JAVA(100, "Java"),
    C(101, "C"),
    PYTHON(102, "Python"),
    CPP(103, "C++"),
    VB(104, "VB"),
    GO(105, "Go");
    private int index;
    private String desc;
    PLanguage(int id, String desc){
        this.index = id;
        this.desc = desc;
    }
    public int getIndex(){
        return index;
    }
    public String getDesc(){
        return desc;
    }
}
```

第四步：新增枚举转换类 EnumConversion，提供根据枚举字段值查找枚举值的功能。

代码 8-6：EnumConversion 类

```java
public class EnumConversion<E> {
    private String methodName;
    public EnumConversion(String sMethod){
        this.methodName = sMethod;
    }
    public EnumConversion(){
        this("getIndex");
    }
    public E convertIndexToEnum(Integer index, Class<E> clazz){
        E[ ] enumConstants = clazz.getEnumConstants();
        try {
            Method getEnumID = clazz.getMethod(this.methodName);
            for (E item : enumConstants) {
                Object invokeID = getEnumID.invoke(item);
                if (invokeID.equals(index)) {
                    return item;
                }
            }
        }catch (Exception ex){
            ex.printStackTrace();
```

```
        }
        return null;
    }
    public String convertEnumToIndexString(E eum, Class<E> clazz){
        try {
            Method getEnumID = clazz.getMethod(this.methodName);
            Integer index = (Integer) getEnumID.invoke(eum);
            return String.valueOf(index);
        }catch (Exception ex){
            ex.printStackTrace();
        }
        return "";
    }
}
```

第五步：新增几个自定义的实现 IConvertor 接口的转换器类。

代码 8-7：GenderConvertor 类

```
public class GenderConvertor extends EnumConversion<Gender>
                     implements IConvertor<Gender>{
    @Override
    public Gender convert(String sValue) {
        Integer iValue = Integer.valueOf(sValue);
        return this.convertIndexToEnum(iValue, Gender.class);
    }
}
```

代码 8-8：MajorConvertor 类

```
public class MajorConvertor extends EnumConversion<Major>
                     implements IConvertor<Major>{
    @Override
    public Major convert(String sValue) {
        Integer iValue = Integer.valueOf(sValue);
        return this.convertIndexToEnum(iValue, Major.class);
    }
}
```

代码 8-9：PLanguageConvertor 类

```
public class PLanguageConvertor extends EnumConversion<PLanguage>
                     implements IConvertor<PLanguage> {
    public PLanguageConvertor(String sMethod){
        super(sMethod);
    }
    @Override
    public PLanguage convert(String sValue) {
        Integer iValue = Integer.valueOf(sValue);
```

```
        return this.convertIndexToEnum(iValue, PLanguage.class);
    }
}
```

第六步：添加工具类 CommonUtil。

代码 8-10：CommonUtil 类

```
public class CommonUtil {
    private static final AtomicInteger globalIdentity = new AtomicInteger(100);
    public static String colorToHexString(Color color){
        int r = color.getRed();
        int g = color.getGreen();
        int b = color.getBlue();
        return String.format("#%02X%02X%02X", r, g, b);
    }
    public static Integer nextGlobalInteger(){
        return globalIdentity.getAndIncrement();
    }
}
```

第七步：新增实体类 User，用@FormField 注解对 User 类的属性进行标注。

代码 8-11：User 类

```
@Data
@NoArgsConstructor
@AllArgsConstructor
public class User {
    @FormField(name = "id")
    private Integer userid;

    @FormField
    private String account;

    @FormField
    private String password;

    @FormField(name = "f_username")
    private String name;

    @FormField(name = "gender", convertorClass = GenderConvertor.class)
    private Gender sex;

    @FormField(name = "f_height")
    private double height;

    @FormField(name = "f_weight")
```

```java
    private Integer weight;

    @FormField(convertParams = {"yyyy-MM-dd"})
    private LocalDate birthday;

    @FormField
    private Color color;

    public String getColorString(){
        return CommonUtil.colorToHexString(color);
    }

    @FormField(convertorClass = PLanguageConvertor.class,
                    convertParams = {"getIndex"})
    private List<PLanguage> languages;

    @FormField(name = "major")
    private Major speciality;

    @FormField(name = "email")
    private String mail;

    @FormField
    private String address;

    private LocalDateTime creatime;
}
```

第八步:添加工具类 ConvertorFactory。

代码 8-12:ConvertorFactory 类

```java
public class ConvertorFactory {
    private static Map<String, Class<?>> primitiveToWrapper = Map.of(
            "boolean", Boolean.class,
            "byte", Byte.class,
            "char", Character.class,
            "short", Short.class,
            "int", Integer.class,
            "long", Long.class,
            "float", Float.class,
            "double", Double.class
    );
    //若@FormField 未指明转换器类,则根据属性类型从 convertors 中查找转换器
    private static Map<Class<?>, BiFunction<String, String[], Object>>
                            convertors = new HashMap<>();
```

```java
static {
    //往convertors中添加LocalDate类型属性的转换器
    convertors.put(LocalDate.class, (x, y)-> {
        try {
            return LocalDate.parse(x, DateTimeFormatter.ofPattern(y[0]));
        }catch (Exception e){
            //e.printStackTrace();
        }
        return null;
    });
    //往convertors中添加LocalDateTime类型属性的转换器
    convertors.put(LocalDateTime.class, (x, y)-> LocalDateTime.parse(x,
            DateTimeFormatter.ofPattern(y[0])));
    //往convertors中添加Date类型属性的转换器
    convertors.put(Date.class, (x, y)-> {
        SimpleDateFormat simpleDateFormat = new SimpleDateFormat(y[0]);
        Date date = null;
        try{
            date = simpleDateFormat.parse(x);
        }catch (Exception e){
            //e.printStackTrace();
        }
        return date;
    });
    //往convertors中添加Boolean类型属性的转换器
    convertors.put(Boolean.class, (x,y)-> Boolean.parseBoolean(x));
    //往convertors中添加Byte类型属性的转换器
    convertors.put(Byte.class, (x,y)-> Byte.parseByte(x));
    //往convertors中添加Character类型属性的转换器
    convertors.put(Character.class, (x, y)-> Character.valueOf(x.charAt(0)));
    //往convertors中添加Short类型属性的转换器
    convertors.put(Short.class, (x,y)-> Short.parseShort(x));
    //往convertors中添加Integer类型属性的转换器
    convertors.put(Integer.class, (x,y)-> {
        try{
            return Integer.parseInt(x);
        }catch (Exception ex){
            return null;
        }
    });
    //往convertors中添加Long类型属性的转换器
    convertors.put(Long.class, (x,y)-> Long.parseLong(x));
```

```java
        //往convertors中添加Float类型属性的转换器
        convertors.put(Float.class, (x,y)-> Float.parseFloat(x));
        //往convertors中添加Double类型属性的转换器
        convertors.put(Double.class, (x,y)-> {
            double value = 0.0;
            try {
                value = Double.parseDouble(x);
            }catch (Exception e){
                //e.printStackTrace();
            }
            return value;
        });
        //往convertors中添加String类型属性的转换器
        convertors.put(String.class, (x,y)-> x);
        //往convertors中添加Major类型属性的转换器
        convertors.put(Major.class, (x, y) -> new EnumConversion<Major>().
                    convertIndexToEnum(Integer.valueOf(x),Major.class));
        //往convertors中添加Color类型属性的转换器
        convertors.put(Color.class, (x,y)-> Color.decode("0x" +x.substring(1)));
    }
    //根据指定的IConvertor转换器类及构造参数，创建转换器实例
    private static IConvertor customConvertor(Class<? extends IConvertor>
                            convertorClazz, String[ ] convertParams){
        //若指定的转换器类为IConvertor（即默认值），则不创建转换器实例
        //这表示将从convertors里查找转换器
        if (convertorClazz == IConvertor.class){
            return null;
        }
        //根据构造参数生成参数类型数组，以便随后选择构造器创建实例
        Class<String>[ ] paramTypes = null;
        if (convertParams!=null && convertParams.length>0){
            paramTypes = new Class[convertParams.length];
            for(int i=0;i<paramTypes.length;i++){
                paramTypes[i] = String.class;
            }
        }
        IConvertor convertor = null;
        try {
            //若没有构造参数
            if(paramTypes==null){
                //则使用无参构造器创建转换器实例
                convertor = convertorClazz.getDeclaredConstructor().newInstance();
```

```java
        }else {
            //否则使用有参构造器创建转换器实例
            convertor = convertorClazz.getDeclaredConstructor(paramTypes)
                            .newInstance(convertParams);
        }
    }catch (Exception ex){
        ex.printStackTrace();
    }
    return convertor;
}
//map 表示 ServletRequest 的参数名及其对应的多个值(一个参数名可对应多个值)
//此函数将 ServletRequest 的参数名及其值转换成目标类型的对象
//目标类型的@FormField 注解决定哪些是需要转换的参数名及其值
public static Object mapToObject(Map<String, String[ ]> map,
                        Class<?> targetClass) throws Exception{
    if (map == null){ return null;}
    //创建目标类型的对象,此时对象的属性值为初始值
    Object target = targetClass.getConstructor().newInstance();
    //获取目标类型的所有属性
    Field[ ] fields = target.getClass().getDeclaredFields();
    //对每个属性分别处理
    for(Field field : fields){
        //获取当前属性的修饰符
        int mod = field.getModifiers();
        //忽略 static 或 final 属性
        if (Modifier.isStatic(mod) || Modifier.isFinal(mod)){
            continue;
        }
        //使得属性信息是可访问的
        field.setAccessible(true);
        //获取当前属性的@FormField 注解
        FormField formField = field.getAnnotation(FormField.class);
        //如果当前属性不存在@FormField 注解,则忽略此属性
        if (formField == null) { continue; }
        //获取@FormField 注解的 name 属性值作为 ServletRequest 的参数名
        String paramName = formField.name();
        //如果@FormField 注解的 name 属性值为空(即默认值)
        if (paramName.isBlank()){
            //则把当前属性名作为 ServletRequest 的参数名
            paramName = field.getName();
        }
        //从 ServletRequest 获取参数名 paramName 所对应的多值
```

```java
String[ ] paramValues = map.get(paramName);
//若参数名 paramName 对应的多值为 null 则忽略此属性
if (paramValues == null) {
    continue;
}
//获取@FormField 注解的 convertParams 属性值
String[ ] convertParams = formField.convertParams();
//根据@FormField 注解指定的转换器类及构造参数，创建转换器实例
IConvertor customConvertor =
    customConvertor(formField.convertorClass(), convertParams);
Class<?> fieldType;
//如果当前属性的类型是 List
if(field.getType().equals(List.class)){
    List list = new ArrayList();
    //获取 List<T>泛型的参数化类型
    ParameterizedType listGenericType =
                (ParameterizedType) field.getGenericType();
    //获取泛型在运行时的实际类型
    //数组表示泛型可能有多个，而 List<T>只有一个泛型
    Type[ ] listActualTypeArguments =
            listGenericType.getActualTypeArguments();
    //获取此 List 类型属性运行时列表元素类型
    Class genericType =
        Class.forName(listActualTypeArguments[0].getTypeName());
    //从 convertors 查找列表元素类型转换器
    BiFunction<String, String[ ], Object> f = convertors.get(genericType);
    //把 ServletRequest 中此参数名对应的多个值分别转换成列表元素
    for(String paramValue: paramValues){
        Object itemValue;
        //若@FormField 指定了自定义转换器
        if (customConvertor!=null){
            //则使用@FormField 中指定的转换器进行转换
            itemValue = customConvertor.convert(paramValue);
        }else {
            //否则使用从 convertors 查找到的转换器进行转换
            itemValue = f.apply(paramValue, convertParams);
        }
        //每次转换一个值，将转换结果存入列表
        list.add(itemValue);
    }
    //将生成的列表赋值给目标对象的当前属性
    field.set(target, list);
```

```java
    //继续处理目标对象的下一个属性
    continue;
}
//如果当前属性的类型是数组
if (field.getType().isArray()){
    //获取数组元素类型
    Class componentType = field.getType().getComponentType();
    //根据此参数值的个数，创建等长的数组
    Object array =
            Array.newInstance(componentType, paramValues.length);
    int index = 0;
    //从convertors中查找数组元素类型转换器
    BiFunction<String, String[ ], Object> g =
                            convertors.get(componentType);
    //把ServletRequest中此参数名对应的多个值分别转换成数组元素
    for(String paramValue: paramValues){
        Object itemValue;
        //若@FormField指定了自定义转换器
        if (customConvertor!=null){
            //则使用@FormField中指定的转换器进行转换
            itemValue = customConvertor.convert(paramValue);
        }else {
            //否则使用从convertors中查找到的转换器进行转换
            itemValue = g.apply(paramValue, convertParams);
        }
        //每次转换一个值，将转换结果存入数组
        Array.set(array, index++, itemValue);
    }
    //将生成的数组赋值给目标对象的当前属性
    field.set(target, array);
    //继续处理目标对象的下一个属性
    continue;
}
//若当前属性是基本类型
if(field.getType().isPrimitive()){
    //则找到与基本类型对应的封装器类型，以便后续查找转换器
    fieldType = primitiveToWrapper.get(field.getType().toString());
}else{
    //否则是一般类型（即非数组、非列表、非基本类型）
    fieldType = field.getType();
}
//从convertors中查找类型转换器
```

```
            BiFunction<String, String[ ], Object> fun = convertors.get(fieldType);
            Object fieldValue;
            //若@FormField指定了自定义转换器
            if (customConvertor!=null){
                //则使用@FormField中指定的转换器进行转换
                fieldValue = customConvertor.convert(paramValues[0]);
            }else {
                //否则使用从convertors中查找到的转换器进行转换
                fieldValue = fun.apply(paramValues[0], convertParams);
            }
            //将转换结果赋值给目标对象的当前属性
            field.set(target, fieldValue);
        }
        //返回目标类型的实例
        return target;
    }
}
```

第九步：添加 UserServletRequestListener 类，它将用户提交的信息转换成 User 对象。

代码 8-13：UserServletRequestListener 类

```
public class UserServletRequestListener implements ServletRequestListener {
    @Override
    public void requestInitialized(ServletRequestEvent sre) {
        HttpServletRequest req = (HttpServletRequest)sre.getServletRequest();
        try {
            //当前requestInitialized(.)方法在Filter之前执行，此时Filter中的
            //req.setCharacterEncoding(.)方法尚未生效，故在此处设置字符编码
            req.setCharacterEncoding("utf-8");
        }catch (Exception e){ }
        if (req.getServletPath().equals("/user")){
            if (req.getMethod().equals("POST")){
                User user = null;
                Map<String, String[ ]> parameters = req.getParameterMap();
                try {
                    //将表单数据转换成User对象
                    user = (User) ConvertorFactory.mapToObject(
                                        parameters, User.class);
                    user.setCreatime(LocalDateTime.now());
                }catch (Exception e){
                    e.printStackTrace();
                }
                //将User对象存入ServletRequest的命名属性
```

```
                req.setAttribute(UserServlet.doPost_Parameter_User_ATTR, user);
            }
        }
    }
}
```

第十步：在 ServletInitializer 中向 Servlet 容器注册 UserServletRequestListener 监听器。

代码 8-14：注册 UserServletRequestListener

```
//注册 UserServletRequestListener 监听器
ServletRequestListener userServletRequestListener;
userServletRequestListener = new UserServletRequestListener();
servletContext.addListener(userServletRequestListener);
```

第十一步：新增 IRepository 接口及其实现类 UserRepositoryMem，用 Map 模拟数据库存储。

代码 8-15：IRepository 接口

```
public interface IRepository<T, ID> {
    public T save(T t);
    public void deleteById(ID id);
    public T getById(ID id);
    public List<T> getAll();
}
```

代码 8-16：UserRepositoryMem 类

```
public class UserRepositoryMem implements IRepository<User, Integer> {
    //用 HashMap 模拟数据库存储
    private static final Map<Integer, User> users = new ConcurrentHashMap<>();
    @Override
    public User save(User user) {
        if (user.getUserid()==0){
            user.setUserid(CommonUtil.nextGlobalInteger());
        }
        users.put(user.getUserid(), user);
        return user;
    }
    @Override
    public void deleteById(Integer id) {
        users.remove(id);
    }
    @Override
    public User getById(Integer id) {
        return users.get(id);
    }
    @Override
    public List<User> getAll() {
```

```
        List<User> list = new ArrayList<>(users.values().stream().toList());
        Collections.sort(list, Comparator.comparing(User::getUserid).reversed());
        return list;
    }
}
```

第十二步：新增工具类 RepositoryUtil。

代码 8-17：RepositoryUtil 类

```
public class RepositoryUtil {
    private static final String USER_REPOSITORY_ATTR =
                    "cn.shanghai.cxiao.user.RepositoryInstance";
    public static IRepository<User, Integer> userRepository(
                                ServletContext servletContext){
        return (IRepository<User, Integer>)
                servletContext.getAttribute(USER_REPOSITORY_ATTR);
    }
    public static void initializeRepository(ServletContext servletContext){
        IRepository<User, Integer> userRepository = new UserRepositoryMem();
        //将 UserRepositoryMem 对象存入 ServletContext 的命名属性
        servletContext.setAttribute(USER_REPOSITORY_ATTR, userRepository);
    }
}
```

第十三步：在 GlobalObjectSetup 监听器中增加对 IRepository 的初始化。

代码 8-18：GlobalObjectSetup 类

```
public class GlobalObjectSetup implements ServletContextListener {
    @Override
    public void contextInitialized(ServletContextEvent sce) {
        ServletContext servletContext = sce.getServletContext();
        //创建全局的 TemplateEngine 对象并存入 ServletContext
        ThymeleafUtil.initializeTemplateEngine(servletContext);
        //创建全局的 IRepository 对象并存入 ServletContext
        RepositoryUtil.initializeRepository(servletContext);
    }
}
```

第十四步：添加 UserServlet 类。

代码 8-19：UserServlet 类

```
public class UserServlet extends HttpServlet {
    public static final String doPost_Parameter_User_ATTR =
                    "UserServlet.doPost.Parameter.user";
    @Override
    protected void doGet(HttpServletRequest req, HttpServletResponse resp)
```

```java
        throws ServletException, IOException {
    PLanguage[ ] languages = PLanguage.values();
    Major[ ] majors = Major.values();
    Gender[ ] genders = Gender.values();
    IRepository<User, Integer> userRepository =
            RepositoryUtil.userRepository(req.getServletContext());
    List<User> users = userRepository.getAll();
    String uid = req.getParameter("uid");
    User user;
    if (uid == null){
        user = new User();
        user.setUserid(0);
        user.setColor(new Color(0,0,0));
    }else{
        user = userRepository.getById(Integer.valueOf(uid));
    }
    Map<String, Object> map = Map.of("languages", languages,
            "majors", majors,
            "genders", genders,
            "users", users,
            "user", user
    );
    ThymeleafUtil.outputHTML(req, resp, map, "user/user_edit");
}
@Override
protected void doPost(HttpServletRequest req, HttpServletResponse resp)
throws ServletException, IOException {
//从 ServletRequest 命名属性获取 User 对象
    User user = (User)req.getAttribute(UserServlet.doPost_Parameter_User_ATTR);
    if (user!=null){
        //从 ServletContext 获取 IRepository 对象
        IRepository userRepository =
            RepositoryUtil.userRepository(req.getServletContext());
        //将 User 对象存入模拟的数据库
        userRepository.save(user);
    }
    //跳转到信息录入功能
    resp.sendRedirect("/user");
    }
}
```

第十五步：向 Servlet 容器注册 UserServlet。

代码 8-20：注册 UserServlet

```
//注册 UserServlet
servlet = new UserServlet();
registration = servletContext.addServlet("user.servlet", servlet);
registration.setLoadOnStartup(5);
registration.addMapping("/user/*");
registration.setAsyncSupported(true);
```

第十六步：添加 CSS 文件、图片文件及网页 user_edit.html。下面列出若干页面元素的 Thymeleaf 表示。

代码 8-21：HTML 模板网页中用 Thymeleaf 显示对象

```
<label class="col-2 text-right">性别</label>
<p class="col-4 m-0 form-control">
    <label th:each="gender : ${genders}">
        <input th:name="gender" type="radio" id="gender_2"
            th:value="${gender.getIndex()}"
            th:checked="${user.sex==gender}">
        <span th:text="${gender.getDesc(#locale)}"/>
    </label>
</p>
```

```
<label class="col-2 text-right">生日</label>
<input th:name="birthday" type="date" class="col-4" id="birthday"
    th:value="${#temporals.format(user.birthday, 'yyyy-MM-dd')}"
    max="2010-12-31" min="1980-01-01" value="2000-3-25">
```

```
<label class="text-right col-2">喜好颜色</label>
<input th:name="color" id="color" type="color" class="col-4"
    th:value="${user.getColorString()}">
```

```
<label class="col-2 text-right">熟悉语言</label>
<p class="col-10 m-0">
    <label th:each="language : ${languages}">
        <input type="checkbox" th:name="languages"
```

```
            th:checked="${user.languages!=null &&
                         user.languages.contains(language)}"
            th:value="${language.getIndex()}" id="languages_0">
        <span th:text="${language.getDesc()}"/>
    </label>
</p>
```

专业 　数学

```
<label class="col-2 text-right">专业</label>
<select th:name="major" class="col-4" id="major">
    <option th:each="major:${majors}" th:value="${major.getIndex()}"
            th:selected="${user.speciality==major}"
            th:text="${major.getDesc(#locale)}" />
</select>
```

- Lee
- zhangsan

```
<ul th:each="user:${users}">
    <li><a th:href="@{/user(uid=${user.userid})}"
           th:text="${user.account}" /></li>
</ul>
```

第十七步：运行程序，之后在浏览器地址栏中输入 http://localhost:8080/user，查看显示结果。

8.4　用Spring实现

设计思路

Spring 中无须开发者自己进行输入数据绑定，但是对于非常规的 Java 类型，依然需要开发者提供 Converter 的实现类，该类可以将字符串转换成特定类型的对象。在网页中显示对象时，Spring 使用 Model 对象在 Controller 与网页模板之间传递数据，网页模板内使用 Thymeleaf 表达式获取对象的属性值。实现步骤如下：

第一步：新增若干枚举类，如 Gender、Major、PLanguage，与 8.3 节第三步相同。

第二步：新增枚举转换类 EnumConversion，与 8.3 节第四步相同。

第三步：新增实现 Converter 接口的枚举转换器类 EnumConvertor。

代码 8-22：EnumConvertor 类

```
public class EnumConvertor<T> extends EnumConversion<T>
                    implements Converter<String, T>{
    private Class<T> clazz;
```

```java
public EnumConvertor(Class<T> clazz){
    super();
    this.clazz = clazz;
}
//将 index 文本转换成对应的 Enum 对象
public T convert(String source) {
    Integer index = Integer.valueOf(source);
    return this.convertIndexToEnum(index, this.clazz);
}
```

第四步：新增若干枚举转换器类，如 GenderConvertor、MajorConvertor、PLanguageConvertor。

代码 8-23：GenderConvertor 类

```java
public class GenderConvertor extends EnumConvertor<Gender> {
    public GenderConvertor(){
        super(Gender.class);
    }
}
```

代码 8-24：MajorConvertor 类

```java
public class MajorConvertor extends EnumConvertor<Major> {
    public MajorConvertor(){
        super(Major.class);
    }
}
```

代码 8-25：PLanguageConvertor 类

```java
public class PLanguageConvertor extends EnumConvertor<PLanguage> {
    public PLanguageConvertor(){
        super(PLanguage.class);
    }
}
```

第五步：新增转换器类 ColorConvertor、LocalDateConvertor。

代码 8-26：ColorConvertor 类

```java
public class ColorConvertor implements Converter<String, Color> {
    @Override
    public Color convert(String source) {
        try{
            return Color.decode("0x" +source.substring(1));
        }catch (Exception e){
        }
        return new Color(0,0,0);
    }
}
```

代码 8-27：LocalDateConvertor 类

```java
public class LocalDateConvertor implements Converter<String, LocalDate> {
    @Override
    public LocalDate convert(String source) {
        try {
            return LocalDate.parse(source,
                            DateTimeFormatter.ofPattern("yyyy-MM-dd"));
        }catch (Exception e){
        }
        return null;
    }
}
```

第六步：在 MvcConfig 中重载 addFormatters(.)方法，添加转换器实例。

代码 8-28：addFormatters(.)方法

```java
@Override
public void addFormatters(FormatterRegistry registry) {
    registry.addConverter(new GenderConvertor());
    registry.addConverter(new MajorConvertor());
    registry.addConverter(new PLanguageConvertor());
    registry.addConverter(new LocalDateConvertor());
    registry.addConverter(new ColorConvertor());
}
```

第七步：新增实体类 User，无须添加@FormField 注解。

代码 8-29：User 类

```java
@Data
@NoArgsConstructor
@AllArgsConstructor
public class User {
    private Integer userid;
    private String account;
    private String password;
    private String name;
    private Gender sex;
    private double height;
    private Integer weight;
    private LocalDate birthday;
    private Color color;
    private List<PLanguage> languages;
    private Major speciality;
    private String mail;
    private String address;
```

```
    private LocalDateTime creatime;
}
```

第八步：新增工具类 CommonUtil，与 8.3 节第六步相同。

第九步：新增接口 IRepository 及其实现类 UserRepositoryMem 模拟数据库操作，与 8.3 节第十一步相同，但增加注解@Repository。

第十步：新增 UserController。

代码 8-30：UserController 类

```
@Controller
@RequestMapping("/user")
public class UserController {
    @Autowired
    IRepository<User, Integer> userRepository;
    @GetMapping
    public String setupForm(@RequestParam("uid") @Nullable Integer id,Model model){
        List<User> users = userRepository.getAll();
        User user;
        if (id==null){
            user = new User();
            user.setUserid(0);
            user.setColor(new Color(0,0,0));
        }else{
            user = userRepository.getById(id);
        }
        model.addAttribute("users", users);
        model.addAttribute("user", user);
        return "user/user_edit";
    }
    @PostMapping
    public String submitForm(User user){
        user.setCreatime(LocalDateTime.now());
        userRepository.save(user);
        return "redirect:/user";
    }
}
```

第十一步：添加 CSS 文件、图片文件及网页 user_edit.html。网页中表单输入控件的名称必须与 User 的属性名相同。下面列出若干页面元素的 Thymeleaf 表示。

代码 8-31：HTML 模板网页中用 Thymeleaf 显示对象

```
<label class="col-2 text-right">性别</label>
```

```html
<p class="col-4 m-0 form-control">
    <label th:each="gender : ${T(cxiao.sh.cn.enumeration.Gender).values()}">
        <input th:name="sex" type="radio" id="gender_2"
                th:value="${gender.getIndex()}"
                th:checked="${user.sex==gender}">
            <span th:text="${gender.getDesc(#locale)}"/>
    </label>
</p>
```

生日 　2010/12/09　📅

```html
<label class="col-2 text-right">生日</label>
<input th:name="birthday" type="date" class="col-4" id="birthday"
        th:value="${#temporals.format(user.birthday, 'yyyy-MM-dd')}"
        max="2010-12-31" min="1980-01-01" value="2000-3-25">
```

喜好颜色　▇▇▇▇▇▇

```html
<label class="text-right col-2">喜好颜色</label>
<input th:name="color" id="color" type="color" class="col-4"
        th:value="${T(cxiao.sh.cn.utils.CommonUtil)
        .colorToHexString(user.color)}">
```

熟悉语言　☐Java ☑C ☐Python ☑C++ ☑VB ☐Go

```html
<label class="col-2 text-right">熟悉语言</label>
<p class="col-10 m-0">
    <label th:each="language : ${T(cxiao.sh.cn.enumeration.PLanguage).values()}">
        <input type="checkbox" th:name="languages" th:checked=
            "${user.languages!=null && user.languages.contains(language)}"
            th:value="${language.getIndex()}" id="languages_0">
        <span th:text="${language.getDesc()}"/>
    </label>
</p>
```

专业　　数学　　　　　　　　　∨

```html
<label class="col-2 text-right">专业</label>
<select th:name="speciality" class="col-4" id="major">
    <option th:each="major:${T(cxiao.sh.cn.enumeration.Major).values()}"
            th:value="${major.getIndex()}"
            th:selected="${user.speciality ==major}"
            th:text="${major.getDesc(#locale)}" />
</select>
```

```html
<ul th:each="user:${users}">
  <li><a th:href="@{/user(uid=${user.userid})}" th:text="${user.account}" />
  </li>
</ul>
```

第十二步：在 ServletInitializer 类的 onStartup(.)方法中向 Servlet 容器注册 CharacterEncodingFilter，解决表单提交时中文乱码问题。

代码 8-32：注册 CharacterEncodingFilter

```
CharacterEncodingFilter filter = new CharacterEncodingFilter("utf-8", true);
FilterRegistration filterRegistration =
                servletContext.addFilter("encodingFilter", filter);
filterRegistration.addMappingForUrlPatterns(null, true, "/*");
```

第十三步：运行程序，之后在浏览器地址栏中输入 http://localhost:8081/user，查看显示结果。

8.5　用SpringBoot实现

设计思路

与 Spring 实现方案的设计思路相同。实现步骤如下：

第一步：新增若干枚举类，如 Gender、Major、PLanguage，与 8.4 节第一步相同。

第二步：新增枚举转换类 EnumConversion，与 8.4 节第二步相同。

第三步：新增实现 Converter 接口的枚举转换类 EnumConvertor，与 8.4 节第三步相同。

第四步：新增若干枚举转换类，如 GenderConvertor、MajorConvertor、PLanguageConvertor，与 8.4 节第四步相同。

第五步：新增转换类 ColorConvertor、LocalDateConvertor，与 8.4 节第五步相同。

第六步：在 MvcConfig 中重载 addFormatters(.)方法，添加转换器实例，与 8.4 节第六步相同。

第七步：新增实体类 User，与 8.4 节第七步相同。

第八步：新增工具类 CommonUtil，与 8.4 节第八步相同。

第九步：新增 IRepository 接口及其实现类 UserRepositoryMem，用 Map 模拟数据库操作，与 8.4 节第九步相同。

第十步：新增 UserController，与 8.4 节第十步相同。

第十一步：添加 CSS 文件、图片文件及网页 user_edit.html，与 8.4 节第十一步相同。

第十二步：运行程序，之后在浏览器地址栏中输入 http://localhost:8082/user，查看显示结果。

8.6 小结

在进行输入数据绑定时，使用反射机制将文本数据转换成对象属性值，用自定义的 @FormField 标注需要绑定的对象属性并指明转换器类，在 ServletRequestListener 的 requestInitialized(.)方法中完成输入数据绑定。在进行输出数据绑定时，用 Map 在 Servlet 与网页模板之间传递对象，在网页模板内通过 Map 的键名访问对应的对象，并通过 Thymeleaf 表达式访问对象的属性值。

8.7 习题

为本章案例添加一个登录功能，只有输入指定的账号和密码时才能使用用户信息录入及编辑功能。浏览器内登录界面如图 8-5 所示。

图8-5　用户登录界面

第 9 章 文件上传及下载

本章介绍文件上传及下载。本章案例是在录入或者修改用户信息时增加文件上传的功能，以及提供文件下载的功能。

通过学习本章内容，读者将可以：
- 上传文件
- 下载文件

9.1 相关概念

提交表单时，如果需要支持文件上传，那么必须修改<form>的 enctype 属性值。enctype 的不同取值表示表单提交时数据编码格式的不同。

9.1.1 表单的enctype属性

表单 form 的 enctype 属性用来控制浏览器提交表单数据时如何进行编码。enctype 有三种取值，分别为：

- application/x-www-form-urlencoded

浏览器在发送表单数据之前，对所有字符进行编码。此时表单中的文件域只会发送用户选择的文件名，而不会发送文件内容。

application/x-www-form-urlencoded 为默认的编码类型。

- multipart/form-data

不对字符编码，可发送二进制文件。其他两种类型不能用于发送文件。

- text/plain

用于发送纯文本内容，不对特殊字符进行编码，一般用于 email 之类的内容。

9.1.2 Multipart

Multipart 指多部分对象集合，可以容纳多种类型的数据。假设有如图 9-1 所示界面。

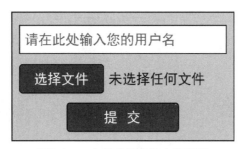

图9-1 一个简单的文件上传界面

当在图 9-1 中填写了用户名 admin，选择了文件 abc.png，单击"提交"按钮后，浏览器向服务器发送的消息头中有如下一行：

```
Content-Type: multipart/form-data; boundary=ZaY03x
```

boundary=ZaY03x 中的 ZaY03x 是随机生成的一段字符串，实际应用中产生的随机字符串可能是这样的：----WebKitFormBoundaryyb1zYhTI38xpQxBK。

浏览器向服务器发送的消息体如下：

```
--ZaY03x
Content-Disposition: form-data; name="username"

admin
--ZaY03x
Content-Disposition: form-data; name="upfile"; filename="abc.png"
Content-Type: image/png

PNG...(abc.png 的内容)...
--ZaY03x--
```

每两个--ZaY03x 之间是 form 的一个键值对，最后以--ZaY03x--表示结尾，也有以--ZaY03x 表示结尾的。

9.2 案例描述

图 9-2 所示为上传界面，图 9-3 所示为下载界面。单击下载界面中的"删除"链接则删除对应的附件。

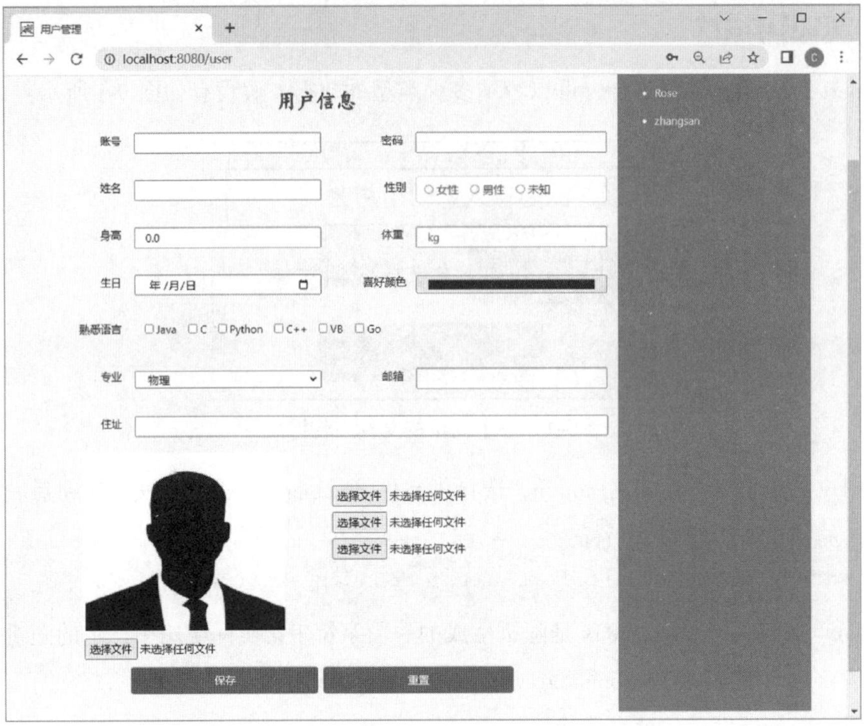

图9-2　案例运行界面（1）

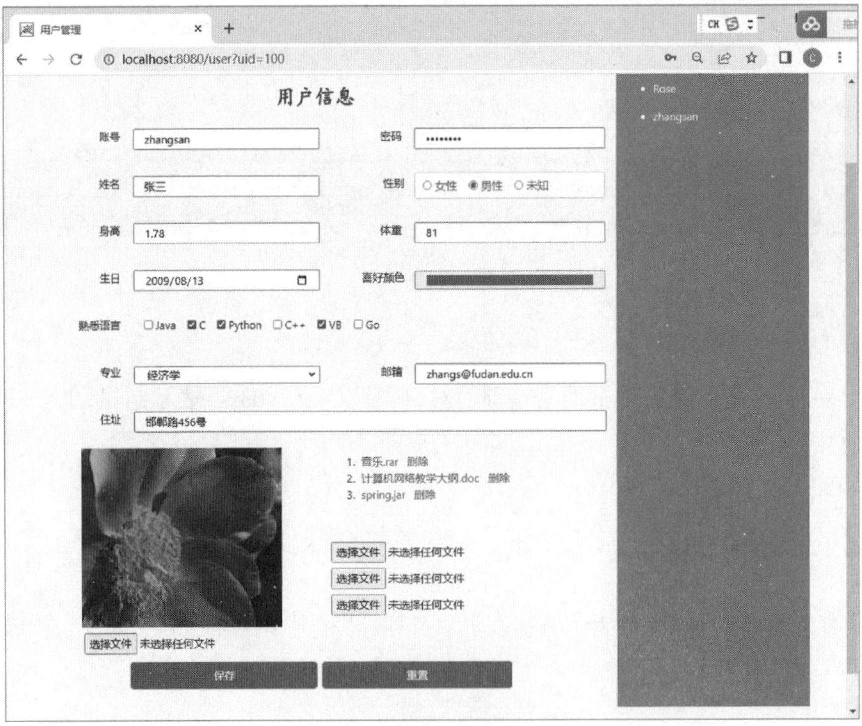

图9-3　案例运行界面（2）

9.3 用Servlet实现

设计思路

创建 MultipartParser 类，将浏览器提交的 Multipart 数据解析成 Map 数据和 Part 数据，其中 Map 存储非上传文件的键值对，而 Part 存储上传文件的内容及文件名等信息。再调用第 8 章的 ConvertorFactory.mapToObject(.)方法将 Map 数据转换成 POJO 对象，而 Part 数据在 POJO 对象写入模拟数据库时保存至文件系统。实现步骤如下：

第一步：添加 AttachmentInfo 类，用于描述附件信息。

代码 9-1：AttachmentInfo 类

```
@Data
@NoArgsConstructor
@AllArgsConstructor
public class AttachmentInfo {
    private String uploadedFileName;
    private String savedFileName;
}
```

第二步：修改 User 类，添加与附件有关的属性。

代码 9-2：User 类中与附件有关的属性

```
private AttachmentInfo photo;
private List<AttachmentInfo> attachments;
```

这两个属性无须添加@FormField 标注，因为不是由表单元素直接转换来的。这两个属性将存放附件上传后的文件名等。

第三步：修改 user_edit.html，增加用于上传的文件域和用于下载的超链接，并修改<form>标签的 enctype 属性为 enctype="multipart/form-data"。

代码 9-3：表单内上传控件和下载链接

```html
<div class="row mx-0 mt-2">
  <div class="col-lg-6 flex-column d-flex align-items-center">
    <img th:src="@{/pic/{filename}(filename=${user.photo.savedFileName})}"
         width="300" height="259" alt=""/>
    <input name="photo" type="file" class="mt-2" id="photo">
  </div>
  <div class="col-lg-6 pt-2">
    <ol th:if="*{attachments!=null}">
      <li th:each="attach:*{attachments}"> 
        <a th:href="@{/user/download(uid=*{userid},
           filename=${attach.savedFileName})}"
           th:text="${attach.uploadedFileName}" />
        <a th:href="@{/user/delete(uid=*{userid},
           filename=${attach.savedFileName})}">
```

```html
            删除</a></li>
        </ol>
        <p> </p>
        <input name="attachments" type="file" class="mb-2" id="attachments1">
        <input name="attachments" type="file" class="mb-2" id="attachments2">
        <input name="attachments" type="file" class="mb-2" id="attachments3">
    </div>
</div>
```

第四步：新增 MultipartParser 类，将请求信息分解成参数名、参数值及 Parts，这里实现了 tomcat 的 UploadContext 接口。

代码 9-4：MultipartParser 类

```java
public class MultipartParser implements UploadContext {
    private static final File TEMP_Directory = new File("d:/tmp/");
    private HttpServletRequest request;
    private String dstFolder;
    //因为 form 表单参数将要作为 mapToObject(.)的参数
    //所以 parameters 类型必须是 Map<String, String[ ]>
    private Map<String, String[ ]> parameters = new HashMap<>();
    //parts 没有上述类型要求，故采用 Map<String, List<Part>>类型
    private Map<String, List<Part>> parts = new HashMap<>();
    public MultipartParser(HttpServletRequest request, String dstFolderPath)
                                                        throws IOException{
        this.request = request;
        this.dstFolder = dstFolderPath;
        AfterConstruct();
    }
    private void AfterConstruct() throws IOException {
        DiskFileItemFactory factory = new DiskFileItemFactory();
        //当文件大小未超过 1GB 时存于内存
        factory.setSizeThreshold((int)Math.pow(2, 30));
        //当文件大小超出上述 SizeThreshold 时暂存于 TEMP_Directory
        factory.setRepository(TEMP_Directory);
        FileUpload upload = new FileUpload();
        upload.setFileItemFactory(factory);
        upload.setHeaderEncoding(this.getCharacterEncoding());
        List<FileItem> fileItems = upload.parseRequest(this);
        for (FileItem fileItem: fileItems){
            String fieldName = fileItem.getFieldName();
            if(fileItem.isFormField()){     //如果是普通表单字段
                String fieldValue = fileItem.getString(this.getCharacterEncoding());
                if (!parameters.containsKey(fieldName)){
                    parameters.put(fieldName, new String[ ]{fieldValue});
                }else{      //form 内同名表单域有多个
```

```java
                    String[ ] oldValues = parameters.get(fieldName);
                    String[ ] newValues = new String[oldValues.length + 1];
                    for (int i=0;i<oldValues.length;i++){
                        newValues[i] = oldValues[i];
                    }
                    newValues[newValues.length-1] = fieldValue;
                    parameters.put(fieldName, newValues);
                }
            }else{    //如果是文件上传字段(即使文件域没有选择文件,也有fileItem)
                Part part = new ApplicationPart(fileItem, new File(this.dstFolder));
                if(!parts.containsKey(fieldName)){
                    parts.put(fieldName, new ArrayList<>(Arrays.asList(part)));
                }else{  //form内同名文件域有多个
                    parts.get(fieldName).add(part);
                }
            }
        }
    }
}
public Map<String, String[ ]> getParameters() {
    return parameters;
}
public Map<String, List<Part>> getParts() {
    return parts;
}
@Override
public long contentLength() {
    return request.getContentLengthLong();
}
@Override
public String getCharacterEncoding() {
    if (this.request.getCharacterEncoding()==null ||
                    this.request.getCharacterEncoding().equals("")){
        return "utf-8";
    }
    return this.request.getCharacterEncoding();
}
@Override
public String getContentType() {
    return this.request.getContentType();
}
@Override
public InputStream getInputStream() throws IOException {
    return this.request.getInputStream();
}
```

```java
public static AttachmentInfo saveUploadPart(Part part){
    String filename = part.getSubmittedFileName();
    //页面内即使没有选择文件,也会为file控件生成非空part
    //此时part.getSubmittedFileName()==""
    if (filename==""){         //无文件上传
        return null;
    }
    String postFix = filename.substring(filename.lastIndexOf("."));
    String uuidName = UUID.randomUUID() + postFix;
    try {
        //将Part的文件内容以uuidName文件名保存
        //目标文件夹在创建Part时已经指定
        part.write(uuidName);
    }catch (Exception e){
        e.printStackTrace();
    }
    return new AttachmentInfo(filename, uuidName);
}
```

第五步:修改 UserServletRequestListener,将第四步解析出来的参数名、参数值(不含 Parts)用 mapToObject(.)方法转换成 User 对象再存入 ServletRequest 的命名属性,而 Parts 直接存入 ServletRequest 的命名属性。

代码 9-5:UserServletRequestListener 类

```java
public class UserServletRequestListener implements ServletRequestListener {
    @Override
    public void requestInitialized(ServletRequestEvent sre) {
        HttpServletRequest req = (HttpServletRequest)sre.getServletRequest();
        try {
            req.setCharacterEncoding("utf-8");
        }catch (Exception e){ }
        if (req.getRequestURI().startsWith("/user")){
            if (req.getMethod().equals("POST")){
                User user = null;
                Map<String, String[ ]> parameters;
                MultipartParser multipartParser = null;
                try {
                    //或者用 req.getHeader("Content-Type")
                    String contentType = req.getContentType();
                    //用 Content-Type 消息头区分表单是否支持文件上传
                    if (contentType.contains("multipart/form-data")) {
                        //当表单包含文件域时
                        multipartParser = new MultipartParser(req, "d:\\ImageOutside");
                        //Parts 直接存入 req 的命名属性
```

```
                    req.setAttribute(UserServlet.doPost_Parameter_Parts_ATTR,
                                    multipartParser.getParts());
                    parameters = multipartParser.getParameters();
                }else{
                    //当表单不含文件域时
                    parameters = req.getParameterMap();
                }
                //非文件域的表单字段转换成 User 对象
                user = (User) ConvertorFactory.mapToObject(parameters, User.class);
                System.out.println(user);
            }catch (Exception e){
                e.printStackTrace();
            }
            //将 User 对象存入 req 的命名属性
            req.setAttribute(UserServlet.doPost_Parameter_User_ATTR, user);
        }
    }
}
```

第六步：修改 IRepository，增加 deleteAttachmentByFileName(.)方法。

代码 9-6：IRepository 接口

```
public interface IRepository<T, ID> {
    public T save(T t);
    public void deleteById(ID id);
    public T getById(ID id);
    public List<T> getAll();
    public void deleteAttachmentByFileName(ID uid, String fileToDelete);
}
```

第七步：更改 IRepository 的实现类 UserRepositoryMem。

代码 9-7：UserRepositoryMem 类

```
public class UserRepositoryMem implements IRepository<User, Integer> {
    private static final Map<Integer, User> users = new ConcurrentHashMap<>();
    @Override
    public User save(User user) {
        if (user.getUserid()==0){
            user.setUserid(CommonUtil.nextGlobalInteger());
        }
        users.put(user.getUserid(), user);
        return user;
    }
    @Override
    public void deleteById(Integer id) {
```

```
            users.remove(id);
        }
        @Override
        public User getById(Integer id) {
            return users.get(id);
        }
        @Override
        public List<User> getAll() {
            List<User> list = new ArrayList<>(users.values().stream().toList());
            Collections.sort(list, Comparator.comparing(User::getUserid).reversed());
            return list;
        }
        @Override
        public void deleteAttachmentByFileName(Integer uid, String fileToDelete) {
            User user = users.get(uid);
            user.getAttachments().removeIf(x->x.getSavedFileName().equals(fileToDelete));
            if (new File("D:/ImageOutside/" + fileToDelete).delete()) {
                System.out.println("成功删除记录号为" + uid +"的附件:" + fileToDelete);
            }
        }
    }
```

第八步：本章案例不考虑验证，故现在只需要保存 Parts，并将保存的文件名填入 User 对象的 photo 属性或者 attachments 属性。为 UserServlet 新增如下两个方法：

代码 9-8：UserServlet 新增的两个方法

```
public static void transferUploadedToUserInfo(User user,
                            Map<String, List<Part>> parts,
                            IRepository<User, Integer> userRepository){
    //界面上是新增用户
    if(user.getUserid()==0){
        user.setCreatime(LocalDateTime.now());
        //附件信息
        appendAttachmentsToUserInfo(user, parts);
        //照片上传：页面内名为"photo"的 file 控件只有一个
        Part part = parts.get("photo").get(0);
        AttachmentInfo info = MultipartParser.saveUploadPart(part);
        if(info==null){      //无照片上传
            info = new AttachmentInfo(User.DEFAULT_PHOTO_FILENAME,
                            User.DEFAULT_PHOTO_FILENAME);
        }
        user.setPhoto(info);
    }else{               //界面上是修改用户
        User oldUser = userRepository.getById(user.getUserid());
        user.setCreatime(oldUser.getCreatime());
```

```
            //表单不会提交原有附件信息，所以转换出来的 user 不含原有附件信息
            //从 Repository 读取此 user 的原有附件信息
            user.setAttachments(oldUser.getAttachments());
            //把表单提交的新附件信息添加到此 user 的原有附件信息之后
            appendAttachmentsToUserInfo(user, parts);
            //照片上传：页面内名为"photo"的 file 控件只有一个
            Part part = parts.get("photo").get(0);
            AttachmentInfo info = MultipartParser.saveUploadPart(part);
            if(info==null){      //无照片上传
               info = oldUser.getPhoto();
            }
            user.setPhoto(info);
        }
    }
    private static void appendAttachmentsToUserInfo(User user,
                                     Map<String, List<Part>> parts){
        //若有附件上传，则新增的附件会追加至 user.attachments 末尾
        //若无附件上传，则不会更改 user.attachments
        for(Part part:parts.get("attachments")){
            AttachmentInfo info =MultipartParser.saveUploadPart(part);
            if(info!=null){
                List<AttachmentInfo> attaches = user.getAttachments();
                if(attaches==null){
                    attaches = new ArrayList<>();
                }
                attaches.add(info);
                user.setAttachments(attaches);
            }
        }
    }
}
```

第九步：修改 UserServlet 类。

POST 操作时把 User 对象存入数据库，并往数据库中追加附件信息，如果是照片则删除之前的外部文件，再用新照片文件信息更新数据库；Get 操作时若无 uid 参数则新建 User 对象，填充 photo 属性为默认值(此时尚未涉及数据库操作)，若有 uid 参数则查询数据库获取 User 对象。

代码 9-9：UserServlet 类

```
public class UserServlet extends HttpServlet {
    public static final String doPost_Parameter_User_ATTR =
                                "UserServlet.doPost.Parameter.user";
    public static final String doPost_Parameter_Parts_ATTR =
                                "UserServlet.doPost.Parameter.parts";
    @Override
    protected void doGet(HttpServletRequest req, HttpServletResponse resp)
```

```java
        throws ServletException, IOException {
    PLanguage[ ] languages = PLanguage.values();
    Major[ ] majors = Major.values();
    Gender[ ] genders = Gender.values();
    IRepository<User, Integer> userRepository =
            RepositoryUtil.userRepository(req.getServletContext());
    List<User> users = userRepository.getAll();
    String path = req.getPathInfo();
    String uid = req.getParameter("uid");
    String filename = req.getParameter("filename");
    if (path!=null && path.startsWith("/delete")){
        deleteAttachmentByFilename(Integer.valueOf(uid), filename, resp);
        return;
    }
    if (path!=null && path.startsWith("/download")){
        downloadAttachment(Integer.valueOf(uid), filename, req, resp);
        return;
    }
    User user;
    if (uid == null){
        user = new User();
        user.setUserid(0);
        user.setColor(new Color(0,0,0));
        user.setPhoto(new AttachmentInfo(User.DEFAULT_PHOTO_FILENAME,
                        User.DEFAULT_PHOTO_FILENAME));
    }else{
        user = userRepository.getById(Integer.valueOf(uid));
    }
    Map<String, Object> map = Map.of("languages", languages,
            "majors", majors,
            "genders", genders,
            "users", users,
            "user", user
    );
    ThymeleafUtil.outputHTML(req, resp, map, "user/user_edit");
}
@Override
protected void doPost(HttpServletRequest req, HttpServletResponse resp)
        throws ServletException, IOException {
    User user = (User)req.getAttribute(UserServlet.doPost_Parameter_User_ATTR);
    Map<String, List<Part>> parts = (Map<String, List<Part>>)
            req.getAttribute(UserServlet.doPost_Parameter_Parts_ATTR);
    IRepository<User, Integer> userRepository =
            RepositoryUtil.userRepository(this.getServletContext());
```

```java
        transferUploadedToUserInfo(user, parts, userRepository);
        userRepository.save(user);
        resp.sendRedirect("/user");
    }
    public static void transferUploadedToUserInfo(User user,
                            Map<String, List<Part>> parts,
                            IRepository<User, Integer> userRepository){
        //代码见第八步
    }
    private static void appendAttachmentsToUserInfo(User user,
                            Map<String, List<Part>> parts){
        //代码见第八步
    }
    private void deleteAttachmentByFilename(Integer uid,
                            String fileToDelete,
                            HttpServletResponse resp)
                            throws IOException {
        IRepository<User, Integer> userRepository =
                RepositoryUtil.userRepository(this.getServletContext());
        userRepository.deleteAttachmentByFileName(uid, fileToDelete);
        resp.sendRedirect("/user?uid=" + uid);
    }
    private void downloadAttachment(Integer uid, String fileTodownload,
                            HttpServletRequest req,
                            HttpServletResponse resp)
                            throws IOException {
        IRepository<User, Integer> userRepository =
                RepositoryUtil.userRepository(this.getServletContext());
        //查找附件的显示名称
        User user = userRepository.getById(uid);
        String fileNameDisplayed = "";
        for (int i=0;i<user.getAttachments().size();i++){
            if (user.getAttachments().get(i)
                    .getSavedFileName().equals(fileTodownload)){
                fileNameDisplayed =
                    user.getAttachments().get(i).getUploadedFileName();
                break;
            }
        }
        File file = new File("D:/ImageOutside/" + fileTodownload);
        if (!file.exists()) {
            return;
        }
        InputStream is = new FileInputStream(file);
```

```
            resp.setContentType("application/force-download");
            resp.addHeader("Content-Disposition", "attachment;fileName=" +
                        URLEncoder.encode(fileNameDisplayed, "UTF-8"));
            new WriteListenerForResponse(is, req, resp).setup();
        }
    }
```

第十步：运行程序，之后在浏览器地址栏中输入 http://localhost:8080/user，查看显示结果。

9.4 用Spring实现

设计思路

在 Spring 容器中添加 MultipartResolver Bean，再设置 MultipartConfig。UserController 内方法的 Part 类型参数会自动绑定上传文件的内容。文件下载的函数使用 HttpServletResponse 作为参数，从这个参数的输出流把文件内容发送至浏览器。实现步骤如下：

第一步：添加 AttachmentInfo 类，用于描述附件信息。

代码 9-10：AttachmentInfo 类

```
@Data
@NoArgsConstructor
@AllArgsConstructor
public class AttachmentInfo {
    private String uploadedFileName;
    private String savedFileName;
}
```

第二步：修改 User 类，增加与附件相关的两个属性。

代码 9-11：User 新增的两个属性

```
private AttachmentInfo photo;
private List<AttachmentInfo> attachments;
```

第三步：修改 user_edit.html，增加用于上传的文件域和用于下载的超链接，并修改<form>标签的 enctype 属性为 enctype="multipart/form-data"。

由于 Spring 的对象转换是根据页面元素的名称自动对应的，所以这里的<input>的 name 不能与第二步新增的 User 属性名称相同。

代码 9-12：表单内上传控件和下载链接

```
<div class="row mx-0 mt-2">
    <div class="col-lg-6 flex-column d-flex align-items-center">
        <img th:src="@{/pic/{filename}(filename=${user.photo.savedFileName})}"
                width="300" height="259" alt=""/>
        <input name="photo_control" type="file" class="mt-2" id="photo">
    </div>
    <div class="col-lg-6 pt-2">
```

```html
            <ol th:if="*{attachments!=null}">
                <li th:each="attach:*{attachments}"> 
                <a th:href="@{/user/download(uid=*{userid},
                                filename=${attach.savedFileName})}"
                                th:text="${attach.uploadedFileName}" />
                <a th:href="@{/user/delete(uid=*{userid},
                                filename=${attach.savedFileName})}">
                    删除</a></li>
            </ol>
            <p> </p>
            <input name="attachments_control" type="file" id="attachments1">
            <input name="attachments_control" type="file" id="attachments2">
            <input name="attachments_control" type="file" id="attachments3">
    </div>
</div>
```

第四步：在 MvcConfig 类中添加 MultipartResolver Bean。

代码 9-13：MultipartResolver Bean

```java
@Bean
public MultipartResolver multipartResolver() {
    return new StandardServletMultipartResolver();
}
```

第五步：在 ServletInitializer 类中设置 MultipartConfig。

代码 9-14：设置 MultipartConfig

```java
registration.setMultipartConfig(
        new MultipartConfigElement("D:\\ImageOutside"));
```

还可指定其他可选的限制条件，如上传文件的大小限制。

第六步：在 CommonUtil 中添加 saveUploadPart(.)方法。

代码 9-15：saveUploadPart(.)方法

```java
public static AttachmentInfo saveUploadPart(Part part){
    String filename = part.getSubmittedFileName();
    if (filename==""){        //无文件上传
        return null;
    }
    String postFix = filename.substring(filename.lastIndexOf("."));
    String uuidName = UUID.randomUUID() + postFix;
    try {
        part.write(uuidName);
    }catch (Exception e){
        e.printStackTrace();
    }
    return new AttachmentInfo(filename, uuidName);
}
```

第七步：在 UserController 中添加将 Parts 转换成 User 属性值的两个函数。

代码 9-16：UserController 新增的两个方法

```java
public void transferUploadedToUserInfo(User user, Part part, Part[ ] attachments){
    //界面上是新增用户
    if(user.getUserid()==0){
        user.setCreatime(LocalDateTime.now());
        //附件信息
        appendAttachmentsToUserInfo(user, attachments);
        //照片上传
        AttachmentInfo info = CommonUtil.saveUploadPart(part);
        if (info==null){       //无照片上传
            info = new AttachmentInfo(User.DEFAULT_PHOTO_FILENAME,
                            User.DEFAULT_PHOTO_FILENAME);
        }
        user.setPhoto(info);
    }else{                     //界面上是修改用户
        User oldUser = userRepository.getById(user.getUserid());
        user.setCreatime(oldUser.getCreatime());
        //表单不会提交原有附件信息，所以转换出来的 user 不含原有附件信息
        //从 Repository 读取此 user 的原有附件信息
        user.setAttachments(oldUser.getAttachments());
        //把表单提交的新附件信息添加到此 user 的原有附件信息之后
        appendAttachmentsToUserInfo(user, attachments);
        //照片上传
        AttachmentInfo info = CommonUtil.saveUploadPart(part);
        if (info==null){       //无照片上传
            info = oldUser.getPhoto();
        }
        user.setPhoto(info);
    }
}
private void appendAttachmentsToUserInfo(User user, Part[ ] parts){
    //若有附件上传，则新增的附件会追加至 user.attachments 末尾
    //若无附件上传，则不会更改 user.attachments
    for (Part part : parts){
        AttachmentInfo info =CommonUtil.saveUploadPart(part);
        if (info!=null){
            List<AttachmentInfo> attaches = user.getAttachments();
            if (attaches==null){
                attaches = new ArrayList<>();
            }
            attaches.add(info);
```

```
            user.setAttachments(attaches);
        }
    }
}
```

第八步，修改 IRepository 接口，增加 deleteAttachmentByFileName(.)方法，与 9.3 节第六步相同。

第九步：更改 IRepository 的实现类 UserRepositoryMem，与 9.3 节第七步相同。

第十步：修改 UserController 类。

代码 9-17：UserController 类

```
@Controller
@RequestMapping("/user")
public class UserController {
    @Autowired
    IRepository<User, Integer> userRepository;
    @GetMapping
    public String setupForm(@RequestParam("uid") @Nullable Integer id, Model model){
        List<User> users = userRepository.getAll();
        User user;
        if(id==null){
            user = new User();
            user.setUserid(0);
            user.setColor(new Color(0,0,0));
            user.setPhoto(new AttachmentInfo(User.DEFAULT_PHOTO_FILENAME,
                                    User.DEFAULT_PHOTO_FILENAME));
        }else{
            user = userRepository.getById(id);
        }
        model.addAttribute("users", users);
        model.addAttribute("user", user);
        return "user/user_edit";
    }
    @PostMapping
    public String submitForm(User user,
                @RequestPart("photo_control") Part photo,
                @RequestPart("attachments_control") Part[ ] attachments,
                Model model){
        //页面内即使不提交文件，也会为 file 控件生成非空 Part
        //此时 part.getSubmittedFileName()==""
        //因此参数 photo、attachments !=null
        transferUploadedToUserInfo(user, photo, attachments);
        userRepository.save(user);
        System.out.println("保存后的 User: " + user);
```

```java
        return "redirect:/user";
    }
    @GetMapping("/download")
    public void download(@RequestParam("uid") Integer userid,
                @RequestParam("filename") String fileTodownload,
                HttpServletResponse resp) throws IOException {
        //查找附件的显示名称
        User user = userRepository.getById(userid);
        String fileNameDisplayed = "";
        for(int i=0;i<user.getAttachments().size();i++){
            if (user.getAttachments().get(i).getSavedFileName()
                            .equals(fileTodownload)){
                fileNameDisplayed =
                        user.getAttachments().get(i).getUploadedFileName();
                break;
            }
        }
        File file = new File("D:/ImageOutside/" + fileTodownload);
        if(!file.exists()) {
            return ;
        }
        InputStream is = new FileInputStream(file);
        resp.setContentType("application/force-download");
        resp.addHeader("Content-Disposition", "attachment;fileName=" +
                        URLEncoder.encode(fileNameDisplayed, "UTF-8"));
        is.transferTo(resp.getOutputStream());
        is.close();
    }
    @GetMapping("/delete")
    public String delete(@RequestParam("uid") Integer userid,
                @RequestParam("filename") String fileToDelete){
        userRepository.deleteAttachmentByFileName(userid, fileToDelete);
        return "redirect:/user?uid=" + userid;
    }
    public void transferUploadedToUserInfo(User user, Part part, Part[ ] attachments){
        //代码见第七步
    }
    private void appendAttachmentsToUserInfo(User user, Part[ ] parts){
        //代码见第七步
    }
}
```

第十一步：运行程序，之后在浏览器地址栏中输入 http://localhost:8081/user，查看显示结果。

9.5 用SpringBoot实现

设计思路

与 Spring 实现方案的设计思路相同。实现步骤如下：

第一步：添加 AttachmentInfo 类，与 9.4 节第一步相同。

第二步：修改 User 类，增加两个属性，与 9.4 节第二步相同。

第三步：修改 user_edit.html，增加用于上传和下载的页面元素，并修改\<form>标签的 enctype 属性为 enctype="multipart/form-data"，与 9.4 节第三步相同。

第四步：在 application.properties 中添加 Multipart 配置信息。

代码 9-18：Multipart 配置信息

```
# 配置 MultipartConfig
spring.servlet.multipart.enabled=true
spring.servlet.multipart.location=D:/ImageOutside
spring.servlet.multipart.file-size-threshold=1GB
spring.servlet.multipart.max-file-size=2GB
spring.servlet.multipart.max-request-size=4GB
```

第五步：在 CommonUtil 中添加 saveUploadPart(.)方法，与 9.4 节第六步相同。

第六步：在 UserController 中添加将 Parts 转换成 User 属性值的两个函数，与 9.4 节第七步相同。

第七步：修改 IRepository 接口，增加 deleteAttachmentByFileName(.)方法，与 9.4 节第八步相同。

第八步：更改 IRepository 的实现类 UserRepositoryMem，与 9.4 节第九步相同。

第九步：修改 UserController 类，与 9.4 节第十步相同。

第十步：运行程序，之后在浏览器地址栏中输入 http://localhost:8082/user，查看显示结果。

9.6 小结

文件上传需要修改表单的 enctype 属性值，在 UploadContext 接口的自定义实现类中对 Multipart 格式数据进行解析，将其转换成字符串的键值对（键为表单域名，值为表单域值）以及 Part 对象，键值对可以进一步转换成 POJO 对象，而 Part 对象中的上传文件内容可以进一步保存至文件系统。文件下载则使用非阻塞输出将文件内容发送至浏览器。

9.7 习题

为本章案例添加一个下载照片的超链接并实现其功能。

第 10 章
验证用户输入

本章介绍对象验证，根据所配置的验证注解对某个 POJO 对象进行验证，所配置的验证注解可以是预定义注解也可以是自定义注解。本章案例是在录入或者修改用户信息时增加输入验证功能。

通过学习本章内容，读者将可以：

❑ 进行用户输入的验证

10.1 相关概念

在 Web 应用的三层（表示层、业务层、数据层）架构中，每一层都可能进行数据验证而且通常是相同的验证逻辑。如果在每一层中都实现这种验证逻辑，那么不但会延长开发时间，而且会增加冗余代码。为了避免这种验证代码在各层重复，一种考虑是将验证逻辑直接绑定到领域模型上，将 POJO 类与其验证约束紧密关联，这样一来，验证约束实际上成了类本身的元数据。

Jakarta-Bean-Validation 3.0 规范定义了用于验证对象和对象图的 API，也定义了针对构造器参数、针对方法参数和返回值进行验证的 API，其官方实现为 Hibernate-Validator。使用此 Validator 时，可以通过注解的方式在对象模型上表达约束，也可以用扩展的方式编写自定义约束，不论采用何种方式，都可以报告违反约束的集合。

验证的处理流程如图 10-1 所示。

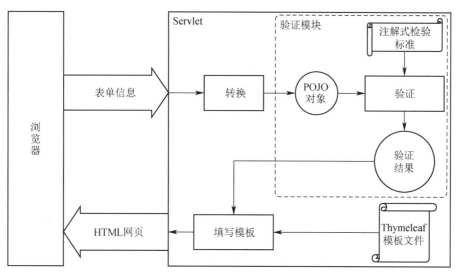

图10-1 用户输入验证的处理流程

10.1.1 预定义的验证注解

Jakarta-Bean-Validation 规范所定义的验证注解见表 10-1，代码中可以直接使用。

表 10-1 预定义的验证注解

注 解	说 明
@Null	限制只能为 null
@NotNull	限制必须不为 null。当 @NotNull 注解被使用在 String 类型的数据上，则表示该数据不能为 null，但是可以为 ""；一个 BigDecimal 的字段应该使用 @NotNull
@NotBlank	限制字符串必须要有非空格的字符。只能用于 String 类型的非空校验，表示该数据不能为 null 且 trim()之后 size>0
@NotEmpty	限制必须不为 null 且不为空。用于 String、Collection、Map、数组等，表示该数据不能为 null 且长度不能为 0
@Size(max,min)	限制（String、Collection、Map、数组等）元素个数必须在 min 到 max 之间
@Length	限制 String 的最大长度
@AssertFalse	限制必须为 false
@AssertTrue	限制必须为 true
@Future	限制必须是一个将来的日期时间
@Past	限制必须是一个过去的日期时间
@DecimalMax(value)	限制必须为一个不大于指定值的数值
@DecimalMin(value)	限制必须为一个不小于指定值的数值
@Digits(integer,fraction)	限制必须为一个小数，且整数部分的位数不能超过 integer，小数部分的位数不能超过 fraction
@Max(value)	限制必须为一个不大于指定值的整数
@Min(value)	限制必须为一个不小于指定值的整数
@Range(min,max)	限制整数的大小范围

续表

注　　解	说　　明
@Pattern(value)	限制字符串必须符合指定的正则表达式
@Email	限制字符串必须是邮箱格式

10.1.2　自定义的验证注解

开发者亦可自定义验证注解。自定义步骤如下：

（1）创建接口 ConstraintValidator 的实现类，如 V。

（2）新增自定义验证注解，如@A。

（3）在 A 的定义上添加注解@Constraint(validatedBy = V.class)。

（4）在实体类的成员变量上添加验证配置@A。

10.2　案例描述

系统对用户输入进行验证。如果未能通过验证，则显示验证出错的提示信息，浏览器显示效果如图 10-2 所示。

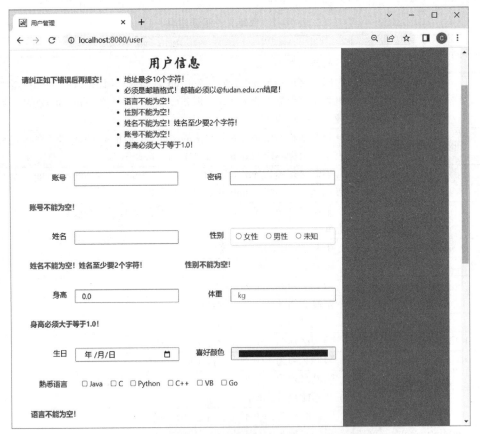

图10-2　案例运行界面

10.3 用Servlet实现

设计思路

根据界面需求创建若干自定义验证注解及其验证器；新增工具类 ValidatorUtil，该工具类可以对具有验证注解的 POJO 对象进行验证，具体的验证工作在过滤器中进行；新增过滤器 UserServletValidateFilter，此过滤器在 UserServlet 处理请求之前对由表单数据转换而来的 User 对象进行验证。如果未通过验证，则将 User 信息及出错提示信息回显至输入页面；如果通过验证，则将 User 信息保存到模拟数据库，之后重定向到新的输入页面。实现步骤如下：

第一步：在模块的 pom.xml 中添加验证相关的依赖。

代码 10-1：与验证相关的依赖

```xml
<dependency>
    <groupId>jakarta.validation</groupId>
    <artifactId>jakarta.validation-api</artifactId>
    <version>3.0.2</version>
</dependency>
<dependency>
    <groupId>org.hibernate.validator</groupId>
    <artifactId>hibernate-validator</artifactId>
    <version>8.0.0.Final</version>
</dependency>
<dependency>
    <groupId>org.apache.tomcat.embed</groupId>
    <artifactId>tomcat-embed-el</artifactId>
    <version>${tomcat.version}</version>
</dependency>
```

第二步：新增 ValidatorUtil 工具类。

代码 10-2：ValidatorUtil 类

```java
public class ValidatorUtil {
    private static Validator validator =
                Validation.buildDefaultValidatorFactory().getValidator();
    public static <T> Map<String, String> validate(T obj){
        return validate(obj, Default.class);
    }
    public static <T> Map<String, String> validate(T obj, Class<?> clazz){
        Map<String, String> errorMap = null;
        Set<ConstraintViolation<T>> violations = validator.validate(obj, clazz);
        if (!violations.isEmpty()){
            String property;
            errorMap = new HashMap<>();
            for(ConstraintViolation<T> constraintViolation : violations){
```

```
                property = constraintViolation.getPropertyPath().toString();
                //一个字段可能经过多次检验，因此同名 property 可能会再次出现
                if(errorMap.get(property)!=null){
                    //追加出错提示信息
                    errorMap.put(property, errorMap.get(property) +
                                    constraintViolation.getMessage());
                }else{
                    //新加出错提示信息
                    errorMap.put(property, constraintViolation.getMessage());
                }
            }
        }
        return errorMap;
    }
}
```

第三步：新增枚举类 StringPattern。

代码 10-3：StringPattern 类

```
package cxiao.sh.cn.enumeration;
public enum StringPattern {
    ALPHABET_DIGIT(1001, "字母和数字"),
    EMAIL_SUFFIX(1002, "邮箱后缀");
    private int index;
    private String desc;
    StringPattern(int index, String desc){
        this.index = index;
        this.desc = desc;
    }
    public int getIndex(){
        return this.index;
    }
    public String getDesc(){
        return this.desc;
    }
}
```

第四步：新增自定义验证注解@SpecialCharacters 的定义。

代码 10-4：SpecialCharacters 注解类

```
@Target({FIELD, ANNOTATION_TYPE})
@Retention(RUNTIME)
//指明此验证注解所用的验证器
@Constraint(validatedBy = SpecialCharactersValidator.class)
public @interface SpecialCharacters {
    //自定义的检验注解
```

```
    String message() default "不符合特定的字符串格式";
    StringPattern pattern();
    Class<?>[ ] groups() default { };
    Class<? extends Payload>[ ] payload() default { };
}
```

第五步:新增@SpecialCharacters 的验证器 SpecialCharactersValidator 的定义。

代码 10-5:SpecialCharactersValidator 类

```
public class SpecialCharactersValidator
            implements ConstraintValidator<SpecialCharacters, String> {
    private SpecialCharacters specialCharacters;
    @Override
    public void initialize(SpecialCharacters constraintAnnotation) {
        this.specialCharacters = constraintAnnotation;
    }
    @Override
    public boolean isValid(String str,
                ConstraintValidatorContext constraintValidatorContext) {
        //对于空串,不作检验。因为空串由@NotBlank 检验
        if (str.equals("")){
            return true;
        }
        if (specialCharacters.pattern().equals(StringPattern.EMAIL_SUFFIX)){
            return str.endsWith("@fudan.edu.cn");
        }
        if (specialCharacters.pattern().equals(StringPattern.ALPHABET_DIGIT)){
            //以字母开头,随后为任意字母或数字
            Pattern pattern = Pattern.compile("[a-zA-Z]+[a-zA-Z0-9]*");
            return pattern.matcher(str).matches();
        }
        return true;
    }
}
```

第六步:增加 User 类各属性的验证标注。

代码 10-6:User 类

```
@Data
@NoArgsConstructor
@AllArgsConstructor
public class User {
    public final static String DEFAULT_PHOTO_FILENAME = "default.PNG";

    @FormField(name = "id")
    private Integer userid;
```

```java
@FormField
@NotBlank(message = "账号不能为空！")
@SpecialCharacters(pattern = StringPattern.ALPHABET_DIGIT ,
            message = "账号必须以字母开头,后接字母或数字！")
private String account;

@FormField
private String password;

@FormField(name = "f_username")
@NotBlank(message = "姓名不能为空！")
@Size(min = 2, message = "姓名至少要2个字符！")
private String name;

@FormField(name = "gender", convertorClass = GenderConvertor.class)
@NotNull(message = "性别不能为空！")
private Gender sex;

@FormField(name = "f_height")
@DecimalMin(value = "1.0", message = "身高必须大于等于1.0！")
@DecimalMax(value = "2.5", message = "身高必须小于等于2.5！")
private double height;

@FormField(name = "f_weight")
@DecimalMin(value = "70", message = "体重必须大于等于70！")
@DecimalMax(value = "120", message = "体重必须小于等于120！")
private Integer weight;

@FormField(convertParams = {"yyyy-MM-dd"})
private LocalDate birthday;

@FormField
private Color color;

public String getColorString(){
    return CommonUtil.colorToHexString(color);
}

@FormField(convertorClass = PLanguageConvertor.class,
           convertParams = {"getIndex"})
@NotEmpty(message = "语言不能为空！")
@Size(min = 2, message = "语言至少选2个！")
private List<PLanguage> languages;
```

```
@FormField(name = "major")
private Major speciality;

@FormField(name = "email")
@Email(message = "必须是邮箱格式！")
@SpecialCharacters(pattern = StringPattern.EMAIL_SUFFIX ,
                   message = "邮箱必须以@fudan.edu.cn 结尾！")
private String mail;

@FormField
@Size(max = 10, message = "地址最多 10 个字符！")
private String address;

private AttachmentInfo photo;
private List<AttachmentInfo> attachments;
private LocalDateTime creatime;
}
```

第七步：新增 UserServletValidateFilter 类，此过滤器对由表单转换来的 User 对象进行验证。

代码 10-7：UserServletValidateFilter 类

```
public class UserServletValidateFilter implements Filter {
    @Override
    public void doFilter(ServletRequest servletRequest,
                     ServletResponse servletResponse,
                     FilterChain filterChain)
                     throws IOException, ServletException {
        HttpServletRequest req = (HttpServletRequest) servletRequest;
        if (req.getMethod().equals("POST")) {
            User user =
                (User)req.getAttribute(UserServlet.doPost_Parameter_User_ATTR);
            if (user==null){
                return;
            }
            //对 User 对象进行验证
            //默认的 Validate Group 是 Default.class
            Map<String, String> errorMap = ValidatorUtil.validate(user);
            if (errorMap!=null){           //当未能通过验证
                PLanguage[ ] languages = PLanguage.values();
                Major[ ] majors = Major.values();
                Gender[ ] genders = Gender.values();
                IRepository<User, Integer> userRepository =
                    RepositoryUtil.userRepository(req.getServletContext());
                List<User> users = userRepository.getAll();
```

```
                if (user.getUserid()!=0){    //修改用户信息时未通过验证
                    User old = userRepository.getById(user.getUserid());
                    if (old!=null) {
                        user.setCreatime(old.getCreatime());
                        user.setPhoto(old.getPhoto());
                        user.setAttachments(old.getAttachments());
                    }
                }else{                        //新增用户信息时未通过验证
                    user.setCreatime(LocalDateTime.now());
                    user.setPhoto(new AttachmentInfo(
                            User.DEFAULT_PHOTO_FILENAME,
                            User.DEFAULT_PHOTO_FILENAME));
                }
                Map<String, Object> map = Map.of("languages", languages,
                        "majors", majors,
                        "genders", genders,
                        "users", users,
                        "user", user,            //回显用户信息
                        "errors", errorMap       //比正常的页面显示增加 errorMap
                );
                ThymeleafUtil.outputHTML(req, (HttpServletResponse)
                        servletResponse, map, "user/user_edit");
                return;
            }
        }
        //若通过验证,则继续过滤器链的处理
        filterChain.doFilter(req, servletResponse);
    }
}
```

第八步：在 ServletInitializer 中向 Servlet 容器注册 UserServletValidateFilter。

代码 10-8：注册 UserServletValidateFilter

```
//此过滤器对录入的用户信息进行验证
filter = new UserServletValidateFilter();
filterRegistration = servletContext.addFilter("userValidateFilter", filter);
filterRegistration.addMappingForServletNames(
    EnumSet.of(DispatcherType.REQUEST), true, "user.servlet");
```

第九步：修改 user_styles.css，增加一项样式规则。

代码 10-9：一条 CSS 样式

```
.error{
    color:#F00;
    font-weight: bold;
}
```

此样式将以红色加粗字体显示出错信息。在第十步会用到此样式。

第十步：修改 user_edit.html，增加各项验证信息展示所用 div，这里按 User 对象属性分条显示。

代码 10-10：HTML 模板网页中显示验证信息

对象属性的验证注解
```
@NotBlank(message = "账号不能为空！")
@SpecialCharacters(pattern = StringPattern.ALPHABET_DIGIT ,
           message = "账号必须以字母开头，后接字母或数字！")
private String account;
```

```
<div th:if="${errors!=null && errors.get('account')!=null}"
       class="row m-0 rowInForm align-items-center error">
    <div class="col-6">
        <span class="col-form-label" th:text="${errors.get('account')}" />
    </div>
</div>
```

对象属性的验证注解
```
@NotBlank(message = "姓名不能为空！")
@Size(min = 2, message = "姓名至少要2个字符！")
private String name;
@NotNull(message = "性别不能为空！")
private Gender sex;
```

```
<div th:if="${errors!=null and (errors.get('name')!=null or
               errors.get('sex')!=null)}"
       class="row m-0 rowInForm align-items-center error">
    <div class="col-6">
        <span class="col-form-label" th:text="${errors.get('name')}" />
    </div>
    <div class="col-6">
        <span class="col-form-label" th:text="${errors.get('sex')}" />
    </div>
</div>
```

对象属性的验证注解
```
@DecimalMin(value = "1.0", message = "身高必须大于等于1.0！")
@DecimalMax(value = "2.5", message = "身高必须小于等于2.5！")
private double height;
@DecimalMin(value = "70", message = "体重必须大于等于70！")
@DecimalMax(value = "120", message = "体重必须小于等于120！")
private Integer weight;
```

```html
<div th:if="${errors!=null and (errors.get('height')!=null or
                      errors.get('weight')!=null)}"
     class="row m-0 rowInForm align-items-center error">
    <div class="col-6">
        <span class="col-form-label" th:text="${errors.get('height')}" />
    </div>
    <div class="col-6">
        <span class="col-form-label" th:text="${errors.get('weight')}" />
    </div>
</div>
```

对象属性的验证注解

```java
@NotEmpty(message = "语言不能为空！")
@Size(min = 2, message = "语言至少选2个！")
private List<PLanguage> languages;
```

```html
<div th:if="${errors!=null and errors.get('languages')!=null}"
             class="row m-0 rowInForm align-items-center error">
    <div class="col-12">
        <span class="col-form-label" th:text="${errors.get('languages')}" />
    </div>
</div>
```

对象属性的验证注解

```java
@Email(message = "必须是邮箱格式！")
@SpecialCharacters(pattern = StringPattern.EMAIL_SUFFIX ,
                   message = "邮箱必须以@fudan.edu.cn结尾！")
private String mail;
```

```html
<div th:if="${errors!=null and errors.get('mail')!=null}"
             class="row m-0 rowInForm align-items-center error">
    <div class="col-6"><span class="col-form-label" /></div>
    <div class="col-6">
        <span class="col-form-label" th:text="${errors.get('mail')}" />
    </div>
</div>
```

对象属性的验证注解

```java
@Size(max = 10, message = "地址最多10个字符！")
private String address;
```

```html
<div th:if="${errors!=null and errors.get('address')!=null}"
             class="row m-0 rowInForm align-items-center error">
  <div class="col-12">
     <span class="col-form-label" th:text="${errors.get('address')}" />
  </div>
</div>
```

```
</div>
```

页面中显示总的出错提示信息

```
<div th:if="${errors!=null and errors.size()>0}" class="d-flex p-0 col-12 mb-3">
    <span class="error">请纠正如下错误后再提交!</span>
    <br>
    <ul>
        <li th:each="errorEntry:${errors}"
            th:text="${errorEntry.value}"></li>
    </ul>
</div>
```

第十一步:运行程序,之后在浏览器地址栏中输入 http://localhost:8080/user,查看显示结果。

10.4 用Spring实现

设计思路

Spring 中可以使用预定义验证注解,亦可使用自定义验证注解。只需在 UserController 内方法的输入绑定参数前标注@Valid 就会由 Spring 框架自动进行验证。在网页模板中用 th:errors 属性显示出错信息。实现步骤如下:

第一步:在模块的 pom.xml 中添加验证相关的依赖。

代码 10-11:与验证相关的依赖

```xml
<dependency>
    <groupId>jakarta.validation</groupId>
    <artifactId>jakarta.validation-api</artifactId>
    <version>3.0.2</version>
</dependency>
<dependency>
    <groupId>org.hibernate.validator</groupId>
    <artifactId>hibernate-validator</artifactId>
    <version>8.0.0.Final</version>
</dependency>
<dependency>
    <groupId>org.apache.tomcat.embed</groupId>
    <artifactId>tomcat-embed-el</artifactId>
    <version>${tomcat.version}</version>
</dependency>
```

第二步:新增枚举类 StringPattern。

代码 10-12:StringPattern 类

```java
package cxiao.sh.cn.enumeration;
public enum StringPattern {
    ALPHABET_DIGIT(1001, "字母和数字"),
```

```
    EMAIL_SUFFIX(1002, "邮箱后缀");
    private int index;
    private String desc;
    StringPattern(int index, String desc){
        this.index = index;
        this.desc = desc;
    }
    public int getIndex(){
        return this.index;
    }
    public String getDesc(){
        return this.desc;
    }
}
```

第三步：新增自定义验证注解@SpecialCharacters 的定义。

代码 10-13：SpecialCharacters 注解类

```
@Target({FIELD, ANNOTATION_TYPE})
@Retention(RUNTIME)
//指明此验证注解所用的验证器
@Constraint(validatedBy = SpecialCharactersValidator.class)
public @interface SpecialCharacters {
    //自定义的检验注解
    String message() default "不符合特定的字符串格式";
    StringPattern pattern();
    Class<?>[ ] groups() default { };
    Class<? extends Payload>[ ] payload() default { };
}
```

第四步：新增@SpecialCharacters 的验证器 SpecialCharactersValidator 的定义。

代码 10-14：SpecialCharactersValidator 类

```
public class SpecialCharactersValidator
                implements ConstraintValidator<SpecialCharacters, String> {
    private SpecialCharacters specialCharacters;
    @Override
    public void initialize(SpecialCharacters constraintAnnotation) {
        this.specialCharacters = constraintAnnotation;
    }
    @Override
    public boolean isValid(String str,
            ConstraintValidatorContext constraintValidatorContext) {
        //对于空串，不作检验。因为空串由@NotBlank 检验
        if (str.equals("")){
            return true;
        }
```

```java
        if (specialCharacters.pattern().equals(StringPattern.EMAIL_SUFFIX)){
            return str.endsWith("@fudan.edu.cn");
        }
        if (specialCharacters.pattern().equals(StringPattern.ALPHABET_DIGIT)){
            //以字母开头，随后任意字母或数字
            Pattern pattern = Pattern.compile("[a-zA-Z]+[a-zA-Z0-9]*");
            return pattern.matcher(str).matches();
        }
        return true;
    }
}
```

第五步：增加 User 类各属性的验证标注。

代码 10-15：User 类

```java
@Data
@NoArgsConstructor
@AllArgsConstructor
public class User {
    public final static String DEFAULT_PHOTO_FILENAME = "default.PNG";

    private Integer userid;

    @NotBlank(message = "账号不能为空！")
    @SpecialCharacters(pattern = StringPattern.ALPHABET_DIGIT ,
                       message = "账号必须以字母开头，后接字母或数字！")
    private String account;

    private String password;

    @NotBlank(message = "姓名不能为空！")
    @Size(min = 2, message = "姓名至少要2个字符！")
    private String name;

    @NotNull(message = "性别不能为空！")
    private Gender sex;

    @DecimalMin(value = "1.0", message = "身高必须大于等于1.0！")
    @DecimalMax(value = "2.5", message = "身高必须小于等于2.5！")
    private double height;

    @DecimalMin(value = "70", message = "体重必须大于等于70！")
    @DecimalMax(value = "120", message = "体重必须小于等于120！")
    private Integer weight;          //可以不填，若填写则必须在指定的大小之内
```

```
private LocalDate birthday;

private Color color;

@NotEmpty(message = "语言不能为空！")
@Size(min = 2, message = "语言至少选2个！")
private List<PLanguage> languages;

private Major speciality;

@Email(message = "必须是邮箱格式！")
@SpecialCharacters(pattern = StringPattern.EMAIL_SUFFIX ,
                   message = "邮箱必须以@fudan.edu.cn结尾！")
private String mail;

@Size(max = 10, message = "地址最多10个字符！")
private String address;

private AttachmentInfo photo;
private List<AttachmentInfo> attachments;
private LocalDateTime creatime;
}
```

第六步：修改 UserController 的 submitForm(.)方法。

因为只有在表单提交时才需要验证，所以这里只修改@PostMapping 所标注的方法。方法的参数 BindingResult result 必须随后紧接@Valid 所标注的参数。

代码 10-16：submitForm(.)方法

```
@PostMapping
public String submitForm(@Valid User user,
                BindingResult result,
                @RequestPart("photo_control") Part photo,
                @RequestPart("attachments_control") Part[ ] attachments,
                Model model){
    if(result.hasErrors()){                    //当未能通过验证
        List<User> users = userRepository.getAll();
        if(user.getUserid()!=0){         //修改用户信息时未通过验证
            User old = userRepository.getById(user.getUserid());
            if(old!=null) {
                user.setCreatime(old.getCreatime());
                user.setPhoto(old.getPhoto());
                user.setAttachments(old.getAttachments());
            }
        }else{                                 //新增用户信息时未通过验证
            user.setCreatime(LocalDateTime.now());
```

```
                user.setPhoto(new AttachmentInfo(User.DEFAULT_PHOTO_FILENAME,
                                         User.DEFAULT_PHOTO_FILENAME));
            }
        model.addAttribute("users", users);
        model.addAttribute("user", user);   //回显用户信息
        return "user/user_edit";
    }
    //当通过了验证
    transferUploadedToUserInfo(user, photo, attachments);
    userRepository.save(user);
    return "redirect:/user";
}
```

第七步：修改 user_styles.css，增加一项样式规则。

代码 10-17：一条 CSS 样式

```
.error{
    color:#F00;
    font-weight: bold;
}
```

此样式将以红色加粗字体显示出错信息。

第八步：修改 user_edit.html，增加各项验证信息展示所用 div，这里按 User 对象属性分条显示。

代码 10-18：HTML 模板网页中显示验证信息

对象属性的验证注解

```
@NotBlank(message = "账号不能为空！")
@SpecialCharacters(pattern = StringPattern.ALPHABET_DIGIT ,
        message = "账号必须以字母开头，后接字母或数字！")
private String account;
```

```
<div th:if="${#fields.hasErrors('account')}"
        class="row m-0 rowInForm align-items-center error">
    <div class="col-6">
        <span class="col-form-label" th:errors="*{account}" />
    </div>
</div>
```

对象属性的验证注解

```
@NotBlank(message = "姓名不能为空！")
@Size(min = 2, message = "姓名至少要2个字符！")
private String name;
@NotNull(message = "性别不能为空！")
private Gender sex;
```

```html
<div th:if="${#fields.hasErrors('name') or #fields.hasErrors('sex')}"
     class="row m-0 rowInForm align-items-center error">
    <div class="col-6">
        <span class="col-form-label" th:errors="*{name}" />
    </div>
    <div class="col-6">
        <span class="col-form-label" th:errors="*{sex}" />
    </div>
</div>
```

对象属性的验证注解
```java
@DecimalMin(value = "1.0", message = "身高必须大于等于1.0！")
@DecimalMax(value = "2.5", message = "身高必须小于等于2.5！")
private double height;
@DecimalMin(value = "70", message = "体重必须大于等于70！")
@DecimalMax(value = "120", message = "体重必须小于等于120！")
private Integer weight;
```

```html
<div th:if="${#fields.hasErrors('height') or
             #fields.hasErrors('weight')}"
     class="row m-0 rowInForm align-items-center error">
    <div class="col-6">
        <span class="col-form-label" th:errors="*{height}" />
    </div>
    <div class="col-6">
        <span class="col-form-label" th:errors="*{weight}" />
    </div>
</div>
```

对象属性的验证注解
```java
@NotEmpty(message = "语言不能为空！")
@Size(min = 2, message = "语言至少选2个！")
private List<PLanguage> languages;
```

```html
<div th:if="${#fields.hasErrors('languages')}"
     class="row m-0 rowInForm align-items-center error">
    <div class="col-12">
        <span class="col-form-label" th:errors="*{languages}" />
    </div>
</div>
```

对象属性的验证注解
```java
@Email(message = "必须是邮箱格式！")
@SpecialCharacters(pattern = StringPattern.EMAIL_SUFFIX ,
                   message = "邮箱必须以@fudan.edu.cn结尾！")
private String mail;
```

```
<div th:if="${#fields.hasErrors('mail')}"
              class="row m-0 rowInForm align-items-center error">
    <div class="col-6"><span class="col-form-label" /></div>
    <div class="col-6">
        <span class="col-form-label" th:errors="*{mail}" />
    </div>
</div>
```

对象属性的验证注解
```
@Size(max = 10, message = "地址最多10个字符！")
private String address;
```

```
<div th:if="${#fields.hasErrors('address')}"
              class="row m-0 rowInForm align-items-center error">
    <div class="col-12">
        <span class="col-form-label" th:errors="*{address}" />
    </div>
</div>
```

页面中显示总的出错提示信息
```
<form method="post" th:object="${user}"  ...... >
    <div th:if="${#fields.hasErrors()}" class="d-flex p-0 col-12 mb-3">
        <span class="error">请纠正如下错误后再提交！</span>
        <br>
        <ul>
            <li th:each="error:${#fields.allErrors()}" th:text= "${error}"></li>
        </ul>
    </div>
    ... ...
</form>
```

第九步：运行程序，之后在浏览器地址栏中输入 http://localhost:8081/user，查看显示结果。

10.5 用SpringBoot实现

设计思路

与 Spring 实现方案的设计思路相同。实现步骤如下：

第一步：在模块的 pom.xml 中添加验证相关的依赖。

代码 10-19：与验证相关的依赖
```
<dependency>
    <groupId>org.springframework.boot</groupId>
    <artifactId>spring-boot-starter-validation</artifactId>
    <version>${springboot.version}</version>
</dependency>
```

第二步：新增枚举类 StringPattern，与 10.4 节第二步相同。

第三步：新增自定义验证注解@SpecialCharacters 的定义，与 10.4 节第三步相同。

第四步：新增@SpecialCharacters 的验证器 SpecialCharactersValidator 的定义，与 10.4 节第四步相同。

第五步：增加 User 类各属性的验证标注，与 10.4 节第五步相同。

第六步：修改 UserController 的 submitForm(.)方法，与 10.4 节第六步相同。

第七步：修改 user_styles.css，增加一项样式规则，与 10.4 节第七步相同。

第八步：修改 user_edit.html，增加各项验证信息展示所用 div，与 10.4 节第八步相同。

第九步：运行程序，之后在浏览器地址栏中输入 http://localhost:8082/user，查看显示结果。

10.6 小结

自定义的验证注解用于标注于类的属性上，与自定义的验证注解搭配使用的是自定义验证器，自定义验证器必须实现 ConstraintValidator 接口。验证操作在过滤器中进行，如果通过了验证（即符合约束），则由 Servlet 继续处理；如果违反了约束，则将出错信息以 Map 对象（键为属性名，值为对应的出错信息）传递至模板网页，最终在浏览器网页中显示。

10.7 习题

为本章案例中"密码"框设置如下约束条件：
- 长度 8~20
- 必须包含大写字母
- 必须包含小写字母
- 必须包含数字
- 必须包含至少一个特殊字符~!@#$%^&*()[ ]{ }<>?\+
- 不能包含空格
- 不能包含上述字符之外的其他字符

当不符合上述约束条件时，则提示密码强度不够。

第 11 章 国 际 化

本章介绍国际化，包括处理请求时读取地区设置、作出应答时指定地区设置，以及根据请求时的语言参数应答对应语言的页面信息、出错提示信息。本章案例是提供不同语言版本的用户信息管理功能。

通过学习本章内容，读者将可以：
- 对页面标签、按钮等文字作国际化
- 对页面列表框、按钮组等内容文字作国际化
- 对出错提示信息作国际化

11.1 相关概念

国际化是提供不同语言版本的软件系统，这并不意味着要开发多个软件系统，而是通过资源属性文件实现的。服务端根据浏览器请求时的地区设置来决定应答时采用何种语言的信息文字。

11.1.1 请求时地区设置

客户端可以选择性地告知 Web 服务器应该采用什么语言作出应答，这个信息在客户端发出请求的消息头 Accept-Language 中指定。ServletRequest 接口的以下方法可以获取客户端首选的地区设置（地区决定了所用语言）。
- getLocale()
- getLocales()

getLocale()方法返回的是客户端首选的地区设置，getLocales()方法返回的是客户端可接受的且按照接受程度从高到低排列的地区设置序列。

如果客户端没有指明地区设置，那么 getLocale()返回的是 Servlet 容器的默认地区设置，而 getLocales()返回的是只有一个元素（即上述默认地区设置）的地区设置序列。

11.1.2 应答时地区设置

Servlet 在应答时应该指明应答的地区设置。Servlet 所指明的地区设置保存在应答消息头 Content-Language 内，再发送至客户端。

ServletResponse 对象的 setLocale(.)方法用于指明地区设置，这个方法可以被重复调用，但是在应答被提交之后的调用无效。如果 Servlet 在应答被提交之前没有指明地区设置，那么 Servlet 容器的默认地区设置将用于应答的地区设置。

11.2 案例描述

图 11-1 所示为英文版的输入验证出错提示界面，图 11-2 所示为英文版的用户详细信息界面。当单击页面中"中文版"文字链接时，页面会切换成中文界面，并且出错提示信息也以中文显示。

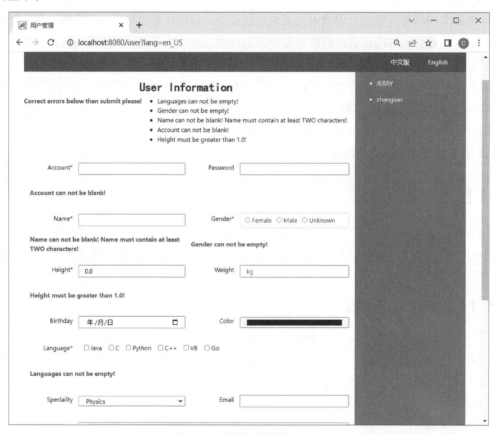

图11-1 案例运行界面（1）

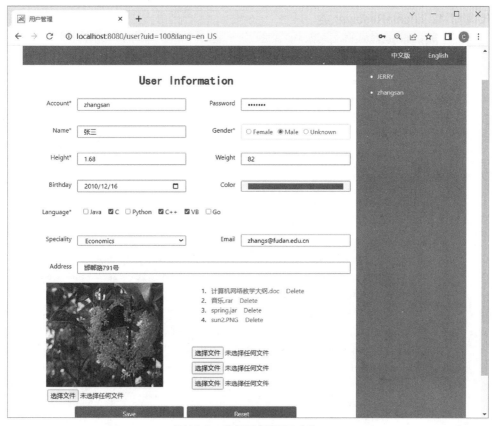

图11-2 案例运行界面（2）

11.3 用Servlet实现

设计思路

在过滤器 UrlJSessionIdFilter（或者是新建的过滤器）中替换 ServletRequest 的 getLocale()方法，使得该方法不再是仅仅从请求消息头 Accept-Language 中获取 Locale 设置，而是优先从请求 URL 的参数 lang 中获取 Locale 设置；只有在请求 URL 中不含参数 lang 时，才从请求消息头 Accept-Language 中获取 Locale 设置。上述替换操作使得在 Servlet 中无须关注 Locale 设置的不同来源，依然是用 req.getLocale()获取 Locale 设置。

添加国际化的出错提示信息文件，在工具类 ValidatorUtil 的验证函数内根据 req.getLocale() 的值选择不同语言的出错提示信息文件并从中读取出错提示信息用于在页面上显示；添加国际化的页面标签、按钮等信息文件，保存在 HTML 网页模板文件相同的文件夹内，之后在 HTML 网页模板文件内用形如 "#{属性名}" 的消息表达式显示页面标签、按钮等文字；枚举类中添加不同语言的描述字段，用于支持页面列表框、按钮组等内容文字的国际化。

实现步骤如下：

第一步：新增 LocaleResolver 类。

代码 11-1：LocaleResolver 类

```java
public class LocaleResolver {
    private static final String LOCALE_PARAM_NAME = "lang";
    private String localeParamName;
    public LocaleResolver(){
        this(LOCALE_PARAM_NAME);
    }
    public LocaleResolver(String paramName){
        this.localeParamName = paramName;
    }
    public String getParamName(){
        return localeParamName;
    }
    public Locale resolveLocale(HttpServletRequest request) {
        String paramValue = request.getParameter(localeParamName);
        System.out.println("lang parameter = " + paramValue);
        Locale locale;
        if (paramValue!=null && paramValue.length()>0) {
            //如果参数不为空，就根据参数值，进行语言切换
            String[ ] s = paramValue.split("_");
            try {
                locale = new Locale(s[0], s[1]);
                //如果拆解正确，但不属于指定的范围时
                if (!locale.equals(Locale.CHINA) && !locale.equals(Locale.US)){
                    locale = Locale.US;
                }
            }catch (Exception e){
                //如果拆解 lang 参数出错，则用.getLocale()代替。
                locale = request.getLocale();
            }
        } else {
            locale = request.getLocale();
        }
        return locale;
    }
}
```

第二步：新增 MyHttpServletRequestWrapper 类，重载 getLocale()方法。

代码 11-2：MyHttpServletRequestWrapper 类

```java
public class MyHttpServletRequestWrapper extends HttpServletRequestWrapper {
    private Locale locale;
    public MyHttpServletRequestWrapper(HttpServletRequest request) {
        super(request);
        LocaleResolver localeResolver = new LocaleResolver();
```

```
            locale = localeResolver.resolveLocale(request);
        }
    @Override
    public Locale getLocale() {
        return this.locale;
    }
}
```

第三步：修改 UrlJSessionIdFilter 类，增加对 MyHttpServletRequestWrapper 的使用。

代码 11-3：UrlJSessionIdFilter 类

```
public class UrlJSessionIdFilter implements Filter {
    @Override
    public void doFilter(ServletRequest servletRequest,
                         ServletResponse servletResponse,
                         FilterChain filterChain)
                    throws IOException, ServletException {
        if (!(servletRequest instanceof HttpServletRequest)) {
            filterChain.doFilter(servletRequest, servletResponse);
            return;
        }
        //把对 Locale 的替换放在这里
        HttpServletRequest httpRequest = (HttpServletRequest) servletRequest;
        MyHttpServletRequestWrapper wrappedRequest =
                        new MyHttpServletRequestWrapper(httpRequest);
        HttpServletResponse httpResponse = (HttpServletResponse) servletResponse;
        HttpServletResponseWrapper wrappedResponse =
                        new HttpServletResponseWrapper(httpResponse){
            @Override
            public String encodeRedirectURL(String url) {
                return url;
            }
            @Override
            public String encodeURL(String url) {
                return url;
            }
        };
        filterChain.doFilter(wrappedRequest, wrappedResponse);
    }
}
```

第四步：新增 MessageResource 类，用于配置国际化信息文件并提供读取功能。

代码 11-4：MessageResource 类

```
public class MessageResource {
    private String basename;
```

```
    Locale locale;
    public MessageResource(String basename, Locale locale){
        this.basename = basename;
        this.locale = locale;
    }
    public MessageResource(String basename){
        this(basename, null);
    }
    public Properties getProperties(){
        String locale_Postfix = "";
        if (locale!=null){
            locale_Postfix = "_" + locale;
        }
        String path = basename + locale_Postfix + ".properties";
        if (!path.startsWith("/")){
            path = "/" + path;
        }
        Properties properties = new Properties();
        InputStream is = this.getClass().getResourceAsStream(path);
        try {
            properties.load(new InputStreamReader(is, "utf-8"));
        }catch (Exception e){
            e.printStackTrace();
        }
        return properties;
    }
}
```

第五步：修改 ValidatorUtil 类，增加生成与 Locale 相关的验证出错信息功能。

代码 11-5：ValidatorUtil 类

```
public class ValidatorUtil {
    private static Validator validator =
            Validation.buildDefaultValidatorFactory().getValidator();
    public static <T> Map<String, String> validate(T obj,
                                MessageResource localeResource){
        return validate(obj, Default.class, localeResource);
    }
    public static <T> Map<String, String> validate(T obj, Class<?> clazz,
                                MessageResource localeResource){
        Map<String, String> errorMap = null;
        //默认的组是 Default.class,此处第二个参数不写也行
        Set<ConstraintViolation<T>> violations = validator.validate(obj, clazz);
        if (!violations.isEmpty()){
            String property;
```

```java
            errorMap = new HashMap<>();
            for(ConstraintViolation<T> constraintViolation : violations){
                property = constraintViolation.getPropertyPath().toString();
                //一个字段可能经过多次检验，因此同名 property 可能会再次出现
                if (errorMap.get(property)!=null){
                    errorMap.put(property, errorMap.get(property) +
                        transformMessageByLocale( constraintViolation.
                        getMessage(), localeResource));
                }else{
                    errorMap.put(property, transformMessageByLocale(
                        constraintViolation.getMessage(), localeResource));
                }
            }
        }
        return errorMap;
    }
    private static String transformMessageByLocale(String message,
                                        MessageResource localeResource){
        Properties properties = localeResource.getProperties();
        Enumeration<String> propertyNames =
                    (Enumeration<String>)properties.propertyNames();
        while(propertyNames.hasMoreElements()){
            String propertyName = propertyNames.nextElement();
            String target = "{" + propertyName + "}";
            message = message.replace(target,
                            (String)properties.get(propertyName));
        }
        return message;
    }
}
```

ValidatorUtil 类的改动需要修改 UserServletValidateFilter 类中的语句：

```java
Map<String, String> errorMap = ValidatorUtil.validate(user,
            new MessageResource("i18n/user", req.getLocale()));
```

第六步：在 resources\i18n\文件夹下新增两个与 Locale 有关的文件，user_en_US.properties 和 user_zh_CN.properties。

代码 11-6：user_en_US.properties

```
user.account.notblank=Account can not be blank!
user.account.specialcharacters=Account must start with alphabet, then alphabet or digit!
user.name.notblank=Name can not be blank!
user.name.size=Name must contain at least TWO characters!
user.sex.notnull=Gender can not be empty!
```

```properties
user.height.notnull=Height can not be empty!
user.height.decimalmin=Height must be greater than 1.0!
user.height.decimalmax=Height must be less than 2.5!
user.weight.decimalmin=Weight must be greater than 70!
user.weight.decimalmax=Weight must less than 120!
user.languages.notempty=Languages can not be empty!
user.languages.size=Languages must contain at least TWO items!
user.mail.email=It must accord with email style!
user.mail.specialcharacters=Email must end with @fudan.edu.cn!
user.address.size=Address contains at most TEN characters!
```

代码 11-7：user_zh_CN.properties

```properties
user.account.notblank=账号不能为空!
user.account.specialcharacters=账号必须以字母开头, 后接字母或数字!
user.name.notblank=姓名不能为空!
user.name.size=姓名至少要2个字符!
user.sex.notnull=性别不能为空!
user.height.notnull=身高不能为空!
user.height.decimalmin=身高必须大于等于1.0!
user.height.decimalmax=身高必须小于等于2.5!
user.weight.decimalmin=体重必须大于等于70!
user.weight.decimalmax=体重必须小于等于120!
user.languages.notempty=语言不能为空!
user.languages.size=语言至少选2个!
user.mail.email=必须是邮箱格式!
user.mail.specialcharacters=邮箱必须以@fudan.edu.cn结尾!
user.address.size=地址最多10个字符!
```

第七步：修改User类中验证出错信息，采用第六步所定义的"{属性名}"方式。

代码 11-8：User 类

```java
@Data
@NoArgsConstructor
@AllArgsConstructor
public class User {
    public final static String DEFAULT_PHOTO_FILENAME = "default.PNG";

    @FormField(name = "id")
    private Integer userid;

    @FormField
    @NotBlank(message = "{user.account.notblank}")
    @SpecialCharacters(pattern = StringPattern.ALPHABET_DIGIT,
                message = "{user.account.specialcharacters}")
    private String account;
```

```java
@FormField
private String password;

@FormField(name = "f_username")
@NotBlank(message = "{user.name.notblank}")
@Size(min = 2, message = "{user.name.size}")
private String name;

@FormField(name = "gender", convertorClass = GenderConvertor.class)
@NotNull(message = "{user.sex.notnull}")
private Gender sex;

@FormField(name = "f_height")
@NotNull(message = "{user.height.notnull}")
@DecimalMin(value = "1.0", message = "{user.height.decimalmin}")
@DecimalMax(value = "2.5", message = "{user.height.decimalmax}")
private double height;

@FormField(name = "f_weight")
@DecimalMin(value = "70", message = "{user.weight.decimalmin}")
@DecimalMax(value = "120", message = "{user.weight.decimalmax}")
private Integer weight;

@FormField(convertParams = {"yyyy-MM-dd"})
private LocalDate birthday;

@FormField
private Color color;

public String getColorString(){
    return CommonUtil.colorToHexString(color);
}

@FormField(convertorClass = PLanguageConvertor.class,
        convertParams = {"getIndex"})
@NotEmpty(message = "{user.languages.notempty}")
@Size(min = 2, message = "{user.languages.size}")
private List<PLanguage> languages;

@FormField(name = "major")
private Major speciality;

@FormField(name = "email")
```

```
        @Email(message = "{user.mail.email}")
        @SpecialCharacters(pattern = StringPattern.EMAIL_SUFFIX ,
                           message = "{user.mail.specialcharacters}")
        private String mail;

        @FormField
        @Size(max = 10, message = "{user.address.size}")
        private String address;

        private AttachmentInfo photo;
        private List<AttachmentInfo> attachments;
        private LocalDateTime creatime;
}
```

至此，验证出错信息的国际化已经完成。

第八步：在 user_edit.html 所在文件夹内新增两个同名的与 Locale 相关的属性文件。

代码 11-9：user_edit_en_US.properties

```
user.info=User Information
user.error=Correct errors below then submit please!
user.account=Account
user.password=Password
user.name=Name
user.gender=Gender
user.height=Height
user.weight=Weight
user.birthday=Birthday
user.color=Color
user.language=Language
user.speciality=Speciality
user.email=Email
user.address=Address
user.ok_button=Save
user.reset_button=Reset
user.delete=Delete
```

代码 11-10：user_edit_zh_CN.properties

```
user.info=用户信息
user.error=请纠正如下错误后再提交！
user.account=账号
user.password=密码
user.name=姓名
user.gender=性别
user.height=身高
user.weight=体重
```

```
user.birthday=生日
user.color=喜好颜色
user.language=熟悉语言
user.speciality=专业
user.email=邮箱
user.address=住址
user.ok_button=保存
user.reset_button=重置
user.delete=删除
```

第九步：用 user_edit_<locale>.properties 文件中的属性名替换 user_edit.html 中对应的内容。

代码 11-11：HTML 网页模板中引用多语言资源属性文件内的属性值

```
大标题
<p class="text-center title m-0" th:text="#{user.info}" />
出错提示信息
<span class="error" th:text="#{user.error}" />
账号标签
<label for="account" class="col-2 text-right" th:text="#{user.account} + '*'" />
密码标签
<label class="col-2 text-right" th:text="#{user.password}" />
姓名标签
<label for="address" class="col-2 text-right" th:text="#{user.name} + '*'" />
性别标签
<label class="col-2 text-right" th:text="#{user.gender} + '*'" />
身高标签
<label for="height" class="col-2 text-right" th:text="#{user.height} + '*'" />
体重标签
<label class="col-2 text-right" th:text="#{user.weight}" />
生日标签
<label class="col-2 text-right" th:text="#{user.birthday}" />
颜色标签
<label class="text-right col-2" th:text="#{user.color}" />
语言标签
<label class="col-2 text-right" th:text="#{user.language} + '*'" />
专业标签
<label class="col-2 text-right" th:text="#{user.speciality}" />
邮箱标签
<label class="col-2 text-right" th:text="#{user.email}" />
住址标签
<label class="col-2 text-right" th:text="#{user.address}" />
附件删除链接
<a th:href="@{/user/delete(uid=*{userid},filename=${attach.savedFileName})}"
           th:text="#{user.delete}" />
保存按钮
```

```html
<input name="submit" type="submit" class="col-4 form-control btn-danger"
       id="submit" th:value="#{user.ok_button}">
```
重置按钮
```html
<input name="reset" type="reset" class="col-4 form-control btn-secondary ml-2"
       id="reset" th:value="#{user.reset_button}">
```

第十步：在 user_edit.html 内添加版本链接，以及给相关链接添加 lang=${#locale}参数。

代码 11-12：版本链接
```html
<a th:href="@{/user}" class="mr-5 text-white" th:title="中文版"
                                              th:text="中文版" />
<a th:href="@{/user(lang='en_US')}" th:title="'English Version'"
                                    th:text="'English'" />
```

同时，在 user_edit.html 内给以下链接添加 lang=${#locale}参数。

代码 11-13：为链接增加语言参数 lang

附件删除链接：
```html
<a th:href="@{/user/delete(uid=*{userid},filename=${attach.savedFileName},
             lang=${#locale})}"  th:text="#{user.delete}" />
```
用户列表中用户账号链接：
```html
<a th:href="@{/user(uid=${user.userid},lang=${#locale})}"
   th:text="${user.account}" />
```

第十一步：user_edit.html 内显示性别、专业等有 Locale 特性的枚举值。

代码 11-14：页面元素显示枚举值

性别枚举值：
```html
<span th:text="${gender.getDesc(#locale)}"/>
```
专业枚举值：
```html
<option th:each="major:${majors}"
        th:value="${major.getIndex()}"
        th:selected="${user.speciality==major}"
        th:text="${major.getDesc(#locale)}" />
```

第十二步：修改 UserServlet 类里的 redirect 链接，添加 lang 参数。

代码 11-15：为 redirect 链接增加语言参数 lang
```java
//resp.sendRedirect("/user?uid=" + uid); 改成如下
resp.sendRedirect("/user?uid=" + uid + "&lang=" + locale);
//resp.sendRedirect("/user"); 改成如下
resp.sendRedirect("/user?lang=" + req.getLocale());
```

第十三步：运行程序，之后在浏览器地址栏中输入 http://localhost:8080/user，查看显示结果。

11.4 用Spring实现

设计思路

在Spring容器中添加LocaleResolver Bean，该Bean优先从请求URL的参数lang中获取Locale设置，只有在请求URL中不含参数lang时，才从请求消息头Accept-Language中获取Locale设置；再创建MessageSource Bean，该Bean指明不同语言的资源属性文件所在位置及名称；最后在HTML网页模板中使用"#{属性名}"引用资源属性文件中该属性名对应的文本值。针对验证出错信息，则创建文件名以ValidationMessages开头的不同语言资源属性文件，然后在类的验证注解内使用"{属性名}"引用 ValidationMessages 资源属性文件中该属性名对应的文本值。实现步骤如下：

第一步：新增实现LocaleResolver 接口的 MyLocaleResolver 类。

代码 11-16：MyLocaleResolver 类

```java
public class MyLocaleResolver implements LocaleResolver {
    private static final String LOCALE_PARAM_NAME = "lang";
    private String localeParamName;
    public MyLocaleResolver(){
        this(LOCALE_PARAM_NAME);
    }
    public MyLocaleResolver(String paramName){
        this.localeParamName = paramName;
    }
    public String getParamName(){
        return localeParamName;
    }
    @Override
    public Locale resolveLocale(HttpServletRequest request) {
        //获取页面手动传递的语言参数值：如zh_CN、en_US、""
        String paramValue = request.getParameter(localeParamName);
        Locale locale;
        if (paramValue!=null && paramValue.length()>0) {
            //如果参数不为空，就根据参数值进行语言切换
            String[ ] s = paramValue.split("_");
            try {
                locale = new Locale(s[0], s[1]);
                //如果拆解正确，但不属于指定的范围时
                if (!locale.equals(Locale.CHINA) && !locale.equals(Locale.US)){
                    locale = Locale.US;
                }
            }catch (Exception e){
                //如果拆解lang参数出错，则用.getLocale()代替
```

```
                locale = request.getLocale();
            }
        } else {
            locale = request.getLocale();
        }
        return locale;
    }
    @Override
    public void setLocale(HttpServletRequest request,
                HttpServletResponse response, Locale locale) {
        throw new UnsupportedOperationException("Cannot change HTTP Accept-
            Language header:use a different locale resolution strategy");
    }
}
```

第二步：在 MvcConfig 类中增加 LocaleResolver Bean，名称必须是 localeResolver。

代码 11-17：LocaleResolver Bean

```
@Bean(name="localeResolver")
public LocaleResolver localeResolver() {
    return new MyLocaleResolver("lang");
}
```

第三步：在 MvcConfig 类中增加 MessageSource Bean。

代码 11-18：MessageSource Bean

```
@Bean
public MessageSource messageSource() {
    ReloadableResourceBundleMessageSource messageSource =
                new ReloadableResourceBundleMessageSource();
    //读取消息文件时的编码格式
    messageSource.setDefaultEncoding("utf-8");
    //消息缓存时间，-1 表示不过期
    messageSource.setCacheMillis(-1);
    //消息文件前缀名
    messageSource.setBasename("classpath:i18n/user");
    messageSource.setFallbackToSystemLocale(true);
    return messageSource;
}
```

第四步：根据第三步配置，在 resources\i18n\文件夹内新增 user_en_US.properties 文件和 user_zh_CN.properties 文件。这两个文件用于配置页面内多语言版本的 Label、按钮等的文字信息。

代码 11-19：user_en_US.properties

```
user.info=User Information
user.error=Correct errors below then submit please!
```

```
user.account=Account
user.password=Password
user.name=Name
user.gender=Gender
user.height=Height
user.weight=Weight
user.birthday=Birthday
user.color=Color
user.language=Language
user.speciality=Speciality
user.email=Email
user.address=Address
user.ok_button=Save
user.reset_button=Reset
user.delete=Delete
```

代码 11-20：user_zh_CN.properties

```
user.info=用户信息
user.error=请纠正如下错误后再提交！
user.account=账号
user.password=密码
user.name=姓名
user.gender=性别
user.height=身高
user.weight=体重
user.birthday=生日
user.color=喜好颜色
user.language=熟悉语言
user.speciality=专业
user.email=邮箱
user.address=住址
user.ok_button=保存
user.reset_button=重置
user.delete=删除
```

第五步：用 user_<locale>.properties 文件中的属性名替换 user_edit.html 中对应的内容。

代码 11-21：HTML 网页模板中引用多语言资源属性文件内的属性值

```
大标题
<p class="text-center title m-0" th:text="#{user.info}" />
出错提示信息
<span class="error" th:text="#{user.error}" />
账号标签
<label for="account" class="col-2 text-right" th:text="#{user.account} + '*'" />
密码标签
```

```html
            <label class="col-2 text-right" th:text="#{user.password}" />
姓名标签
            <label for="address" class="col-2 text-right" th:text="#{user.name} + '*'" />
性别标签
            <label class="col-2 text-right" th:text="#{user.gender} + '*'" />
身高标签
            <label for="height" class="col-2 text-right" th:text="#{user.height} + '*'" />
体重标签
            <label class="col-2 text-right" th:text="#{user.weight}" />
生日标签
            <label class="col-2 text-right" th:text="#{user.birthday}" />
颜色标签
            <label class="text-right col-2" th:text="#{user.color}" />
语言标签
            <label class="col-2 text-right" th:text="#{user.language} + '*'" />
专业标签
            <label class="col-2 text-right" th:text="#{user.speciality}" />
邮箱标签
            <label class="col-2 text-right" th:text="#{user.email}" />
住址标签
            <label class="col-2 text-right" th:text="#{user.address}" />
附件删除链接
            <a th:href="@{/user/delete(uid=*{userid},filename=${attach.savedFileName})}"
                th:text="#{user.delete}" />
保存按钮
            <input name="submit" type="submit" class="col-4 form-control btn-danger"
                id="submit" th:value="#{user.ok_button}">
重置按钮
            <input name="reset" type="reset" class="col-4 form-control btn-secondary ml-2"
                id="reset" th:value="#{user.reset_button}">
```

第六步：在 resources\ 文件夹内添加 ValidationMessages_en_US.properties 文件和 ValidationMessages_zh_CN.properties 文件，用于配置不同语言版本的验证出错信息。

代码 11-22：ValidationMessages_en_US.properties

```
    user.account.notblank=Account can not be blank!
    user.account.specialcharacters=Account must start with alphabet, then
alphabet or digit!
    user.name.notblank=Name can not be blank!
    user.name.size=name must contain at least TWO characters!
    user.sex.notnull=Gender can not be empty!
    user.height.notnull=Height can not be empty!
    user.height.decimalmin=Height must be greater than 1.0!
    user.height.decimalmax=Height must be less than 2.5!
    user.weight.decimalmin=Weight must be greater than 70!
```

```
user.weight.decimalmax=Weight must less than 120!
user.languages.notempty=Languages can not be empty!
user.languages.size=Languages must contain at least TWO items!
user.mail.email=It must accord with email style!
user.mail.specialcharacters=Email must end with @fudan.edu.cn!
user.address.size=Address contains at most TEN characters!
```

代码 11-23：ValidationMessages_zh_CN.properties

```
user.account.notblank=账号不能为空！
user.account.specialcharacters=账号必须以字母开头，后接字母或数字！
user.name.notblank=姓名不能为空！
user.name.size=姓名至少要2个字符！
user.sex.notnull=性别不能为空！
user.height.notnull=身高不能为空！
user.height.decimalmin=身高必须大于等于1.0！
user.height.decimalmax=身高必须小于等于2.5！
user.weight.decimalmin=体重必须大于等于70！
user.weight.decimalmax=体重必须小于等于120！
user.languages.notempty=语言不能为空！
user.languages.size=语言至少选2个！
user.mail.email=必须是邮箱格式！
user.mail.specialcharacters=邮箱必须以@fudan.edu.cn结尾！
user.address.size=地址最多10个字符！
```

第七步：修改 User 类中验证出错信息，采用 ValidationMessages 文件中"{属性名}"方式。

代码 11-24：User 类

```
@Data
@NoArgsConstructor
@AllArgsConstructor
public class User {
    public final static String DEFAULT_PHOTO_FILENAME = "default.PNG";
    private Integer userid;
    @NotBlank(message = "{user.account.notblank}")
    @SpecialCharacters(pattern = StringPattern.ALPHABET_DIGIT ,
                       message = "{user.account.specialcharacters}")
    private String account;
    private String password;

    @NotBlank(message = "{user.name.notblank}")
    @Size(min = 2, message = "{user.name.size}")
    private String name;

    @NotNull(message = "{user.sex.notnull}")
    private Gender sex;
```

```java
    @NotNull(message = "{user.height.notnull}")
    @DecimalMin(value = "1.0", message = "{user.height.decimalmin}")
    @DecimalMax(value = "2.5", message = "{user.height.decimalmax}")
    private double height;

    @DecimalMin(value = "70", message = "{user.weight.decimalmin}")
    @DecimalMax(value = "120", message = "{user.weight.decimalmax}")
    private Integer weight;
    private LocalDate birthday;
    private Color color;

    @NotEmpty(message = "{user.languages.notempty}")
    @Size(min = 2, message = "{user.languages.size}")
    private List<PLanguage> languages;
    private Major speciality;

    @Email(message = "{user.mail.email}")
    @SpecialCharacters(pattern = StringPattern.EMAIL_SUFFIX ,
                    message = "{user.mail.specialcharacters}")
    private String mail;

    @Size(max = 10, message = "{user.address.size}")
    private String address;
    private AttachmentInfo photo;
    private List<AttachmentInfo> attachments;
    private LocalDateTime creatime;
}
```

第八步：在 user_edit.html 内添加版本链接，以及给相关链接添加 lang=${#locale}参数。

代码 11-25：版本链接

```
<a th:href="@{/user}" class="mr-5 text-white" th:title="中文版"
                                              th:text="中文版" />
<a th:href="@{/user(lang='en_US')}" th:title="'English Version'"
                                              th:text="'English'" />
```

同时，在 user_edit.html 内给以下链接添加 lang=${#locale}参数。

代码 11-26：为链接增加语言参数 lang

```
附件删除链接：
<a th:href="@{/user/delete (uid=*{userid},filename=${attach.savedFileName},
        lang=${#locale})}" th:text="#{user.delete}" />
用户列表中用户账号链接：
<a th:href="@{/user(uid=${user.userid},lang=${#locale})}"
    th:text="${user.account}" />
```

因为页面内其他链接已经加了 lang 参数,故删除 user_edit.html 中<form>的属性 th:action="@{/user}";或者修改成 th:action="@{/user(lang=${#locale})}"。

第九步:在 user_edit.html 内显示性别、专业等有 Locale 特性的枚举值。

代码 11-27:页面元素显示枚举值

```
性别枚举值:
<span th:text="${gender.getDesc(#locale)}"/>
专业枚举值:
<option th:each="major:${majors}"
        th:value="${major.getIndex()}"
        th:selected="${user.speciality==major}"
        th:text="${major.getDesc(#locale)}" />
```

第十步:修改 UserController 类中的 redirect 链接,添加 lang 参数。

代码 11-28:为 redirect 链接增加语言参数 lang

```
//return "redirect:/user"; 改成如下:
return "redirect:/user?lang=" + locale;
//return "redirect:/user?uid=" + userid; 改成如下:
return "redirect:/user?uid=" + userid + "&lang=" + locale;
```

第十一步:运行程序,之后在浏览器地址栏中输入 http://localhost:8081/user,查看显示结果。

11.5 用SpringBoot实现

设计思路

与 Spring 实现方案的设计思路相同。实现步骤如下:

第一步:新增 MyLocaleResolver 类,与 11.4 节第一步相同。

第二步:在 MvcConfig 类中增加 LocaleResolver Bean,名称必须是 localeResolver,与 11.4 节第二步相同。

第三步:在 MvcConfig 类中增加 MessageSource Bean,与 11.4 节第三步相同。

第四步:根据第三步配置,在 resources\i18n\文件夹内新增 user_en_US.properties 文件和 user_zh_CN.properties 文件。这两个文件用于配置页面内多语言版本的 Label、按钮等的文字信息。与 11.4 节第四步相同。

第五步:用 user_<locale>.properties 文件中的属性名替换 user_edit.html 中对应的内容,与 11.4 节第五步相同。

第六步:在 resources\文件夹内添加 ValidationMessages_en_US.properties 文件和 ValidationMessages_zh_CN.properties 文件,用于配置不同语言版本的验证出错信息。与 11.4 节第六步相同。

第七步:修改 User 类中验证出错信息,采用"{属性名}"方式。与 11.4 节第七步相同。

第八步:在 user_edit.html 内添加版本链接,以及给相关链接添加 lang=${#locale}参数。与 11.4 节第八步相同。

第九步：在 user_edit.html 内显示性别、专业等有 Locale 特性的枚举值，与 11.4 节第九步相同。

第十步：修改 UserController 类中的 redirect 链接，添加 lang 参数，与 11.4 节第十步相同。

第十一步：运行程序，之后在浏览器地址栏中输入 http://localhost:8082/user，查看显示结果。

11.6 小结

Web 应用中需要提供不同语言文本的地方有三个：页面内标签或按钮等的文字、验证出错提示信息、需要在页面内显示的枚举值。在 Servlet 中这三个地方的国际化需要采用不同的方法，但基本思路是一致的，都是提供不同语言的资源属性文件；区别在于如何指定资源属性文件名称和位置以及如何引用资源属性文件中的属性值。选用何种语言版本则默认是由请求时的地区设置决定，亦可由自定义的 URL 参数决定，在本章案例中，自定义参数的名称为 lang。

11.7 习题

为本章案例"密码"框的验证出错提示信息作国际化，验证约束参见 10.7 节。

第 12 章 JPA

本章介绍 Servlet 如何通过 JPA 操作数据库。本章案例是把操作者在浏览器中录入及修改的用户信息保存至 MySQL 数据库，并把从 MySQL 数据库查询到的用户信息显示在浏览器窗口中。

通过学习本章内容，读者将可以：
- 在 Servlet 中通过 JPA 操作数据库。

12.1 相关概念

Jakarta Persistence 3.1 是一个重量级的规范，它定义了一组接口称为 JPA（Jakarta Persistence API）。JPA 是领域模型的对象域与关系模型的记录域之间的桥梁，其官方实现是 Hibernate-Core。本章案例展示如何用事务把多个单表操作组成级联操作。

12.1.1 JPA概述

JPA 与 JDBC 类似，也是官方定义的一组接口。与 JDBC 不同的是，JPA 是为了实现 ORM（Object-Relational Mapping）而生的，它的作用是在关系型数据库和对象之间形成如下映射，以方便开发者在 Java 应用中管理关系型数据：
- 表格——类
- 行——实例
- 列——属性

JPA 基础概念及对象见表 12-1。

表 12-1　JPA 基础概念及对象

概　　念	对应的 JPA 对象	描　　述
Persistence	Persistence	一个引导类，通过它可以获取一个 EntityManagerFactory 实例
Entity Manager Factory	EntityManagerFactory	被配置的工厂对象，通过它可以获取 EntityManager 实例
Persistence Unit	—	一段具有名称的配置信息，声明了实体类和数据库信息；在创建 EntityManagerFactory 实例时使用
Entity Manager	EntityManager	主要的 API 对象，用来对实体执行操作和查询
Persistence Context	—	某个特定 EntityManager 所管理的实体对象集

从 Persistence 创建 EntityManagerFactory 的示例代码如下，其中参数 EmployeeService 是配置文件中 Persistence Unit 名称：

```
EntityManagerFactory emf = Persistence.createEntityManagerFactory("EmployeeService");
```

Persistence Unit 的描述信息存放于配置文件 resources\META-INF\persistence.xml 内。

从 EntityManagerFactory 创建 EntityManager 的示例代码如下：

```
EntityManager em = emf.createEntityManager();
```

Persistence Context 对象集所维护的持久化对象与表格记录存在映射关系，对持久化对象的新增、更改、删除及查找将最终导致对表格记录的新增、修改、删除及查询。

EntityManage 常用的操作持久化对象的方法有：

- em.persist(.)
- em.detach(.)
- em.find(.)
- em.merge(.)
- em.remove(.)

EntityManager 的以下方法可通过 SQL 语句或者 HQL 语句对表格记录进行增加、删除、修改、查询。

- em.createQuery(.)
- em.createNamedQuery(.)
- em.createNativeQuery(.)

图 12-1 显示了 Jakarta Persistence 各个概念之间的关系。

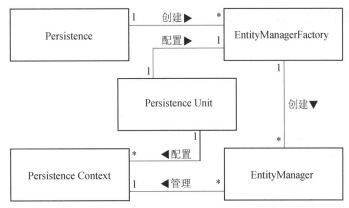

图12-1 Jakarta Persistence各个概念之间的关系

12.1.2 基本注解

JPA 定义了表 12-2 所列的基本注解，用来描述表格与类的映射关系。

表 12-2 JPA 的基本注解

注 解	作 用	举 例
@Entity	表明该类是一个被JPA管理的实体类，该类与数据库一个表格对应	@Entity(name="user") 若表格与实体类同名，则可以不设置 name 值
@Table	当实体类与其映射的数据库表不同名时可以使用 @Table 注解指明数据库表名，该注解与 @Entity 并列使用	@Entity @Table(name="user", schema="servletdb")
@Basic	指明该属性存在与表格列的映射，是一个持久化属性。	若实体类的成员变量上没有任何注解，则默认为@Basic
@Transient	指明该属性不存在与表格列的映射，是一个非持久化属性	@Transient 与@Basic 作用相反
@Column	指明该属性（或 getter()方法）所映射的表格列	@Column(name="gender") 若表格列与实体属性同名，则可以不设置 name 值
@Id	指明该属性与数据库的主键列对应，一个实体类中至少有一个@Id注解	@Id @Column(name="userid") 若表格主键列与实体属性不同名，则增加@Column 注解指明主键列名
@GeneratedValue	定义了主键生成策略。此注解有 strategy 和 generator 两个属性	@GeneratedValue(strategy = GenerationType.IDENTITY) 此例表明主键列采用 ID 自增长方式
@Enumerated	指明表格列与该枚举类型的名称值映射。默认是与该枚举类型的顺序值映射	@Column @Enumerated(EnumType.STRING) private Gender gender; 若不标注@Enumerated，则默认为@Enumerated(EnumType.ORDINAL)

12.1.3 对象的状态转换

图 12-2 显示了 JPA 环境下实体对象的状态转换。

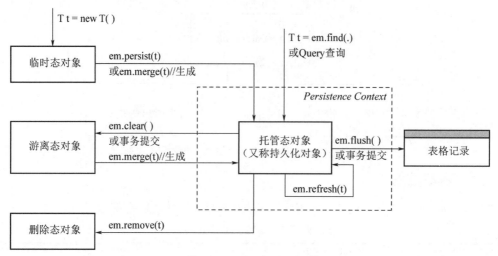

图12-2 实体对象的状态转换图

图 12-2 中四种状态解释如下：

临时态——通过 new 操作创建的实体对象，此对象在表格中没有对应的记录，甚至还未分配主键属性值；该对象与 Persistence Context 没有建立任何关联。

托管态——数据库表格中存在与此对象主键值相同的记录；该对象已经位于 Persistence Context 内，此时对象属性的修改可以直接影响到表格中已有的记录（需要进行事务提交）。

游离态——数据库表格中存在与此对象主键值相同的记录；但是该对象不在 Persistence Context 内，此时对象属性的修改不会影响到表格中已有的记录。

删除态——此对象在表格中没有对应的记录；而且该对象与 Persistence Context 没有关联，因此可以认为删除态是临时态的一种变体。

EntityManager 提供以下四个常用方法，用于实现实体对象的状态转换：

❑ public void persist（Object entity）

persist(.)方法可以将实例由临时态转换为托管态，再调用 flush()方法或提交事务，实例将会被插入到数据库中。对不同状态下的实例 A，persist(.)方法会产生以下结果：

（1）若 A 是临时态，则它将会转为托管态。

（2）若 A 是托管态，则它的状态不会发生任何改变，但是系统仍会在数据库执行 INSERT 操作。

（3）若 A 是删除态，则它将会转换为托管态（当不存在主键冲突时），否则抛出异常。

（4）若 A 是游离态，则该方法会抛出异常。

- public T merge（T var）

merge(.)方法的主要作用是对一个游离态实体的修改进行存库，存库后将产生一个新的托管态对象。对不同状态下的实例 A，merge(.)方法会产生以下结果：

（1）若 A 是游离态，则该方法会将 A 的修改提交到数据库，并返回一个新的托管态实例 A2。

（2）若 A 是临时态，则该方法会生成一个根据 A 产生的托管态新实例 A2。

（3）若 A 是托管态，则它的状态不会发生任何改变，但是系统仍会在数据库执行 UPDATE 操作。

（4）若 A 是删除态，则该方法会抛出异常。

- public void refresh（Object entity）

refresh(.)方法可以保证当前的实例与表格记录的内容一致。对不同状态下的实例 A，refresh(.)方法会产生以下结果：

（1）若 A 是临时态，则不会发生任何操作，但有可能会抛出异常，具体情况根据不同 JPA 实现有关。

（2）若 A 是托管态，则它的属性将会和表格中的记录同步。

（3）若 A 是删除态，则不会发生任何操作。

（4）若 A 是游离态，则该方法会抛出异常。

- public void remove（Object entity）

remove(.)方法可以将实例由托管态转换为删除态，再调用 flush()方法或提交事务，数据库表格中对应的记录将被清除。对不同状态下的实例 A，remove(.)方法会产生以下结果：

（1）如果 A 是临时态，则 A 的状态不会发生任何改变，但系统仍会在数据库中执行 DELETE 语句。

（2）如果 A 是托管态，则它的状态会转换为删除态。

（3）如果 A 是删除态，则不会发生任何操作。

（4）如果 A 是游离态，则该方法会抛出异常。

12.2 案例描述

操作者可以输入信息保存到 MySQL 数据库，操作者亦可从数据库读取所保存的记录。操作者的操作均通过图 12-3 所示的界面进行。

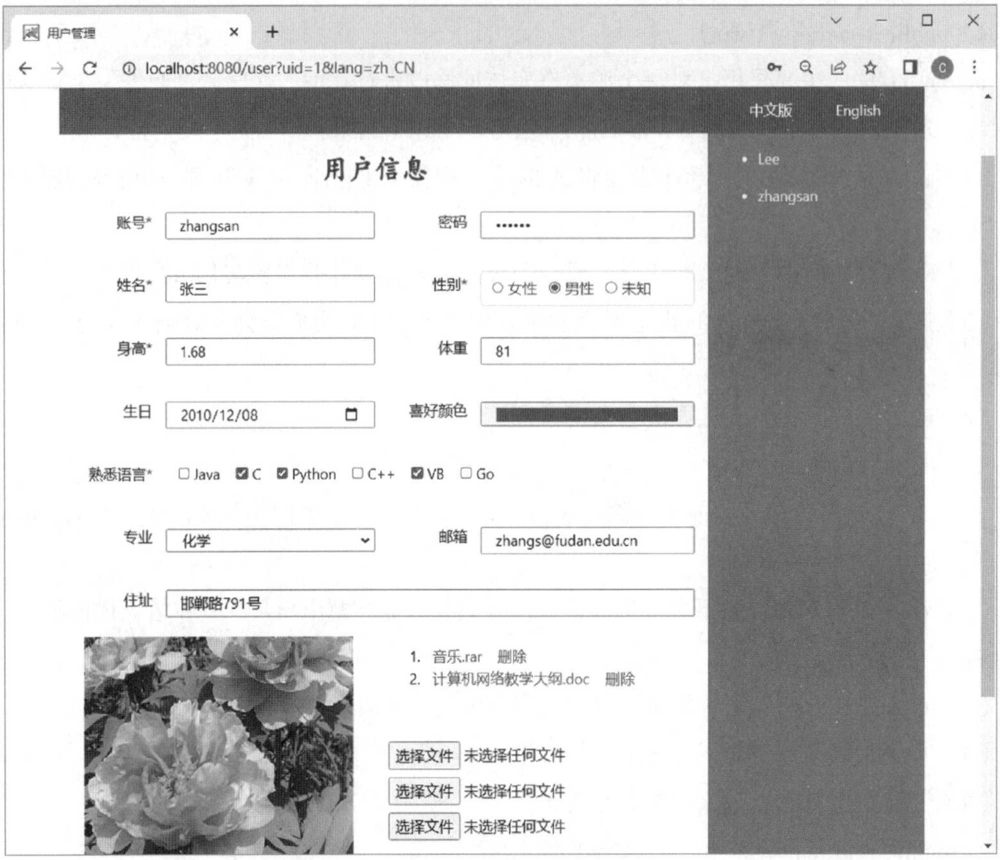

图12-3 案例运行界面

12.3 用Servlet实现

设计思路

配置实体类 User 及其关联类 R_Attachments、R_Photo 及 R_Languages，根据 ORM 的映射关系，这些类将自动创建对应的数据库表格 user、r_attachemnts、r_photo 及 r_languages。新建 IRepository 接口的实现类 UserRepositoryDB，此类提供对上述表格的增删改查操作。

因为数据库是全局对象，所以在 Servlet 容器初始化时创建 UserRepositoryDB 实例并将其存入 ServletContext 的命名属性以便在之后任何时候可以通过此命名属性获取 UserRepositoryDB 对象。

实现步骤如下：

第一步：在模块的 pom.xml 中添加与 Persistence 相关的依赖。

代码 12-1：与 Persistence 相关的依赖

```
<dependency>
    <groupId>jakarta.persistence</groupId>
    <artifactId>jakarta.persistence-api</artifactId>
    <version>3.1.0</version>
</dependency>
```

```xml
<dependency>
    <groupId>org.hibernate.orm</groupId>
    <artifactId>hibernate-core</artifactId>
    <version>6.1.7.Final</version>
</dependency>
<dependency>
    <groupId>mysql</groupId>
    <artifactId>mysql-connector-java</artifactId>
    <version>8.0.30</version>
</dependency>
```

第二步：修改 User 类，增加与 Persistence 相关的注解。

代码 12-2：为 User 类的属性增加相关注解

```
User 类上添加@Entity 注解

@Id
@GeneratedValue(strategy = GenerationType.IDENTITY)
@FormField(name = "id")
private Integer userid;

@Transient
private AttachmentInfo photo;

@Transient
private List<AttachmentInfo> attachments;

@Transient
private List<PLanguage> languages;

//其他成员变量上添加@Column 注解
```

@Transient 表示该属性不参与对象的持久化。

第三步：增加若干与表格对应的实体类，即 R_Attachments、R_Photo、R_Languages，相应的表格用于存储 User 的关联信息。

代码 12-3：R_Attachments 类

```
@Data
@NoArgsConstructor
@AllArgsConstructor
@Entity
@Table(name = "r_attachments")
public class R_Attachments {
    private String uploadedfilename;
    @Id
    private String savedfilename;
```

```
    Integer uid;
}
```

代码 12-4：R_Photo 类

```
@Data
@NoArgsConstructor
@AllArgsConstructor
@Entity
@Table(name = "r_photo")
public class R_Photo {
    private String uploadedfilename;
    private String savedfilename;
    @Id
    Integer uid;
}
```

代码 12-5：R_Languages 类

```
@Data
@NoArgsConstructor
@AllArgsConstructor
@Entity
@Table(name = "r_languages")
public class R_Languages {
    @Id
    Integer uid;
    @Id
    PLanguage language;
}
```

第四步：在 resources\META-INF\文件夹内添加 persistence.xml 文件。

代码 12-6：persistence.xml

```xml
<?xml version="1.0" encoding="UTF-8"?>
<persistence xmlns="http://java.sun.com/xml/ns/persistence"
    xmlns:xsi="http://www.w3.org/2001/XMLSchema-instance"
    xsi:schemaLocation=
            "http://java.sun.com/xml/ns/persistence/persistence_1_0.xsd"
    version="1.0">
    <persistence-unit name="hibernateJPA" transaction-type="RESOURCE_LOCAL">
        <!-- JPA 的实现方式 -->
        <provider>org.hibernate.jpa.HibernatePersistenceProvider</provider>
        <!-- 需要进行 ORM 的 POJO 类，可以有多个 -->
        <class>cxiao.sh.cn.entity.User</class>
        <class>cxiao.sh.cn.entity.R_Attachments</class>
        <class>cxiao.sh.cn.entity.R_Photo</class>
        <class>cxiao.sh.cn.entity.R_Languages</class>
```

```xml
<properties>
    <property name="jakarta.persistence.jdbc.user" value="root"/>
    <property name="jakarta.persistence.jdbc.password"
              value="abcd1234"/>
    <property name="jakarta.persistence.jdbc.driver"
              value="com.mysql.cj.jdbc.Driver"/>
    <property name="jakarta.persistence.jdbc.url"
              value="jdbc:mysql://localhost:3306/servletdb?
                    serverTimezone=Asia/Shanghai
                    &amp;characterEncoding=utf8
                    &amp;useSSL=false"/>
    <property name="hibernate.dialect"
              value="org.hibernate.dialect.MySQLDialect"/>
    <property name="hibernate.hbm2ddl.auto" value="update"/>
    <property name="hibernate.format_sql" value="true"/>
    <property name="hibernate.use_sql_comments" value="true"/>
    <property name="hibernate.show_sql" value="true"/>
</properties>
    </persistence-unit>
</persistence>
```

第五步：新增 UserRepositoryDB 类。

代码 12-7：UserRepositoryDB 类

```java
public class UserRepositoryDB implements IRepository<User, Integer> {
    protected EntityManager em;
    public UserRepositoryDB(EntityManager em){
        this.em = em;
    }
    @Override
    public User save(User user) {
        em.getTransaction().begin();
        if (user.getUserid()==0) {
            user.setUserid(null);
        }
        //savedUser 成为持久化对象
        User savedUser = em.merge(user);
        savedUser.setAttachments(user.getAttachments());
        savedUser.setPhoto(user.getPhoto());
        savedUser.setLanguages(user.getLanguages());
        saveAttachments(savedUser);
        savePhoto(savedUser);
        saveLanguages(savedUser);
        //提交到数据库
        em.getTransaction().commit();
```

```java
            return savedUser;
    }
    private void saveAttachments(User user){
        String executeSQL = "delete from r_attachments where uid = ?";
        em.createNativeQuery(executeSQL).
                setParameter(1, user.getUserid()).executeUpdate();
        if (user.getAttachments()==null) return;
        for (AttachmentInfo info: user.getAttachments()){
            executeSQL = "insert into r_attachments(uid, uploadedfilename, " +
                            "savedfilename) values(?,?,?)";
            em.createNativeQuery(executeSQL)
                    .setParameter(1, user.getUserid())
                    .setParameter(2, info.getUploadedFileName())
                    .setParameter(3, info.getSavedFileName())
                    .executeUpdate();
        }
    }
    private void savePhoto(User user){
        String executeSQL = "delete from r_photo where uid = ?";
        em.createNativeQuery(executeSQL)
                    .setParameter(1, user.getUserid()).executeUpdate();
        if(user.getPhoto()==null) return;
        executeSQL = "insert into r_photo(uid, uploadedfilename, savedfilename) " +
                        "values(?,?,?)";
        em.createNativeQuery(executeSQL)
                .setParameter(1, user.getUserid())
                .setParameter(2, user.getPhoto().getUploadedFileName())
                .setParameter(3, user.getPhoto().getSavedFileName())
                .executeUpdate();
    }
    private void saveLanguages(User user){
        String executeSQL = "delete from r_languages where uid = ?";
        em.createNativeQuery(executeSQL)
                    .setParameter(1, user.getUserid()).executeUpdate();
        if (user.getLanguages()==null) return;
        for (PLanguage language: user.getLanguages()){
            executeSQL = "insert into r_languages(uid, language) values(?,?)";
            em.createNativeQuery(executeSQL)
                    .setParameter(1, user.getUserid())
                    .setParameter(2, language.ordinal())
                    .executeUpdate();
        }
    }
    private void fillTransientFields(User user){
```

```java
        String queryString = "select * from r_attachments where uid = ?";
        List<R_Attachments> result = em
                .createNativeQuery(queryString, R_Attachments.class)
                .setParameter(1, user.getUserid())
                .getResultList();
        if (result!=null && result.size()>0){
            List<AttachmentInfo> list = new ArrayList<>();
            for (R_Attachments attachment: result){
                em.detach(attachment);
                list.add(new AttachmentInfo(attachment.getUploadedfilename(),
                                    attachment.getSavedfilename()));
            }
            user.setAttachments(list);
        }else{
            user.setAttachments(null);
        }
        queryString ="select * from r_photo where uid = ?";
        List<R_Photo> photos = em.createNativeQuery(queryString, R_Photo.class)
                .setParameter(1, user.getUserid())
                .getResultList();
        if (photos!=null && photos.size()>0){
            em.detach(photos.get(0));
            user.setPhoto(new AttachmentInfo(photos.get(0).getUploadedfilename(),
                                    photos.get(0).getSavedfilename()));
        }else{
            user.setPhoto(new AttachmentInfo(User.DEFAULT_PHOTO_FILENAME,
                                    User.DEFAULT_PHOTO_FILENAME));
        }
        queryString = "select * from r_languages where uid = ?";
        List<R_Languages> languages = em.createNativeQuery(queryString,
                                            R_Languages.class)
                                .setParameter(1, user.getUserid())
                                .getResultList();
        if (languages!=null && languages.size()>0){
            List<PLanguage> list = new ArrayList<>();
            for (R_Languages language: languages){
                em.detach(language);
                list.add(language.getLanguage());
            }
            user.setLanguages(list);
        }else{
            user.setLanguages(null);
        }
    }
}
```

```java
@Override
public void deleteById(Integer uid) {
    em.getTransaction().begin();
    User user = em.find(User.class, uid);
    if(user==null){
        return;
    }
    em.remove(user);
    fillTransientFields(user);
    //删除三个关联表
    String sql = "delete from r_attachments where uid = ?";
    em.createNativeQuery(sql)
            .setParameter(1, uid)
            .executeUpdate();
    sql = "delete from r_photo where uid = ?";
    em.createNativeQuery(sql)
            .setParameter(1, uid)
            .executeUpdate();
    sql = "delete from r_languages where uid = ?";
    em.createNativeQuery(sql)
            .setParameter(1, uid)
            .executeUpdate();
    em.getTransaction().commit();
    //删除对应的外部文件
    if(user.getAttachments()!=null) {
        for (AttachmentInfo info : user.getAttachments()) {
            String fileToDelete = info.getSavedFileName();
            if (new File("D:/ImageOutside/" + fileToDelete).delete()) {
                System.out.println("成功删除记录号为" + uid +
                                "的附件:" + fileToDelete);
            }
        }
    }
    if(user.getPhoto()!=null){
        AttachmentInfo info = user.getPhoto();
        String fileToDelete = info.getSavedFileName();
        if(!fileToDelete.equals(User.DEFAULT_PHOTO_FILENAME)) {
            if(new File("D:/ImageOutside/" + fileToDelete).delete()) {
                System.out.println("成功删除记录号为" + uid +
                                "的照片:" + fileToDelete);
            }
        }
    }
}
```

```java
    @Override
    public User getById(Integer uid) {
        User user = em.find(User.class, uid);
        fillTransientFields(user);
        return user;
    }
    @Override
    public List<User> getAll() {
        TypedQuery<User> query = em.createQuery(
                "SELECT u FROM User u order by u.userid desc", User.class);
        List<User> users = query.getResultList();
        return users;
    }
    @Override
    public void deleteAttachmentByFileName(Integer uid, String fileToDelete) {
        String sql =
            "delete from r_attachments where savedfilename = ? and uid = ?";
        em.getTransaction().begin();
        em.createNativeQuery(sql)
                .setParameter(1, fileToDelete)
                .setParameter(2, uid)
                .executeUpdate();
        em.getTransaction().commit();
        if (new File("D:/ImageOutside/" + fileToDelete).delete()) {
            System.out.println("成功删除记录号为" + uid +"的附件:" + fileToDelete);
        }
    }
}
```

第六步：修改 RepositoryUtil 类的 initializeRepository(.)方法。

代码 12-8：RepositoryUtil 类

```java
public class RepositoryUtil {
    private static final String USER_REPOSITORY_ATTR =
                        "cn.shanghai.cxiao.user.RepositoryInstance";
    public static IRepository<User, Integer> userRepository(
                        ServletContext servletContext){
        return (IRepository<User, Integer>)
                servletContext.getAttribute(USER_REPOSITORY_ATTR);
    }
    public static void initializeRepository(ServletContext servletContext){
        EntityManagerFactory emf =
                    Persistence.createEntityManagerFactory("hibernateJPA");
        EntityManager em = emf.createEntityManager();
        em.setFlushMode(FlushModeType.COMMIT);
```

```
            IRepository userRepository = new UserRepositoryDB(em);
            servletContext.setAttribute(USER_REPOSITORY_ATTR, userRepository);
    }
}
```

第七步：修改 user_edit.html，添加"删除"按钮用于删除界面内当前用户。

代码 12-9：添加"删除"按钮

```
<a th:if="*{userid!=0}" th:text="#{user.delete}"
    th:href="@{/user/remove(uid=${user.userid},lang=${#locale})}"
    class="col-3 btn form-control btn-info ml-2"/>
```

第八步：在 UserServlet 中添加删除指定用户的功能。

代码 12-10：添加删除指定用户的功能

```
//先在 UserServlet 的 doGet(.)方法内添加
if (path!=null && path.startsWith("/remove")){
    deleteUser(Integer.valueOf(uid), req.getLocale(), resp);
    return;
}
//然后在 UserServlet 内添加如下函数
private void deleteUser(Integer uid, Locale locale, HttpServletResponse resp)
throws IOException {
    IRepository<User, Integer> userRepository =
                RepositoryUtil.userRepository(this.getServletContext());
    //此函数同时删除了 user 关联的外部文件，default.PNG 除外
    userRepository.deleteById(uid);
    resp.sendRedirect("/user?lang=" + locale);
}
```

第九步：运行程序，之后在浏览器地址栏中输入 http://localhost:8080/user，查看显示结果。

12.4　用Spring实现

设计思路

在 Spring 容器中依次创建存在依赖关系的如下各个 Bean：

- DataSource Bean
- PersistenceUnitManager Bean
- EntityManagerFactory Bean
- PlatformTransactionManager Bean

这些 Bean 会自动完成对 Persistence Context 的初始化；添加若干实体类；新建 UserRepositoryDB 类，用于管理 Persistence Context 内的持久化对象，对应于对表格记录的增删改查操作。实现步骤如下：

第一步：在模块的 pom.xml 中添加 JPA 依赖。

代码 12-11：JPA 依赖

```xml
<dependency>
    <groupId>org.springframework.data</groupId>
    <artifactId>spring-data-jpa</artifactId>
    <version>3.0.3</version>
</dependency>
<dependency>
    <groupId>jakarta.persistence</groupId>
    <artifactId>jakarta.persistence-api</artifactId>
    <version>3.1.0</version>
</dependency>
<dependency>
    <groupId>org.hibernate.orm</groupId>
    <artifactId>hibernate-core</artifactId>
    <version>6.1.7.Final</version>
</dependency>
<dependency>
    <groupId>mysql</groupId>
    <artifactId>mysql-connector-java</artifactId>
    <version>8.0.31</version>
</dependency>
```

第二步：新增 DatabaseConf 类，对数据库进行配置。

代码 12-12：DatabaseConf 类

```java
@Configuration
@EnableJpaRepositories(basePackages = "cxiao.sh.cn.repository")
@EnableTransactionManagement
public class DatabaseConf {
    @Bean
    public Properties hibernateProperties() {
        Properties hibernateProp = new Properties();
        hibernateProp.put("hibernate.dialect", "org.hibernate.dialect.MySQLDialect");
        hibernateProp.put("hibernate.hbm2ddl.auto", "update");
        hibernateProp.put("hibernate.format_sql", false);
        hibernateProp.put("hibernate.use_sql_comments", false);
        hibernateProp.put("hibernate.show_sql", false);
        hibernateProp.put("jakarta.persistence.sharedCache.mode", "NONE");
        return hibernateProp;
    }
    @Bean
    public PersistenceUnitManager persistenceUnitManager(){
        DefaultPersistenceUnitManager persistenceUnitManager =
                                    new DefaultPersistenceUnitManager();
        persistenceUnitManager.setDefaultPersistenceUnitName("myjpa");
```

```java
        //实体类所在的包
        persistenceUnitManager.setPackagesToScan("cxiao.sh.cn.entity");
        persistenceUnitManager.setDefaultDataSource(dataSource());
        return persistenceUnitManager;
    }
    @Bean
    public EntityManagerFactory entityManagerFactory(){
        LocalContainerEntityManagerFactoryBean factoryBean =
                        new LocalContainerEntityManagerFactoryBean();
        factoryBean.setPersistenceUnitManager(persistenceUnitManager());
        factoryBean.setJpaVendorAdapter(new HibernateJpaVendorAdapter());
        factoryBean.setJpaProperties(hibernateProperties());
        factoryBean.afterPropertiesSet();
        return factoryBean.getNativeEntityManagerFactory();
    }
    @Bean
    public PlatformTransactionManager transactionManager() {
        return new JpaTransactionManager(entityManagerFactory());
    }
    @Bean
    public DataSource dataSource() {
        DriverManagerDataSource dataSource = new DriverManagerDataSource();
        dataSource.setDriverClassName(Driver.class.getName());
        dataSource.setUrl("jdbc:mysql://localhost:3306/servletdb?
            serverTimezone=Asia/Shanghai&characterEncoding=utf8&useSSL=false");
        dataSource.setUsername("root");
        dataSource.setPassword("abcd1234");
        return dataSource;
    }
}
```

第三步：给 User 类增加 JPA 注解，与 12.3 节第二步相同。

第四步：新增 R_Attachments 类、R_Photo 类、R_Languages 类，与 12.3 节第三步相同。

第五步：新增 UserRepositoryDB 类，与 12.3 节第五步基本相同，修改三个地方：

在 UserRepositoryDB 类上添加@Repository 注解，删除 UserRepositoryMem 类上的@Repository 注解。

受容器托管的 EntityManager 可以直接通过注解@PersistenceContext 注入的方式获得。

不用 em.getTransaction().begin()开启事务，也不用 em.getTransaction().commit()提交事务，而是在方法上使用 Spring 的@Transactional 注解。

第六步：修改 user_edit.html，添加"删除"按钮用于删除界面内当前用户。

代码 12-13：添加删除按钮

```html
<a th:if="*{userid!=0}" th:text="#{user.delete}"
```

```
th:href="@{/user/remove(uid=${user.userid},lang=${#locale})}"
class="col-3 btn form-control btn-info ml-2"/>
```

第七步：更新 UserController 类。

代码 12-14：UserController 类

```java
@Controller
@RequestMapping("/user")
public class UserController {
    @Autowired
    IRepository<User, Integer> userRepository;
    @GetMapping
    public String setupForm(@RequestParam("uid") @Nullable Integer id,
                         Model model){
        List<User> users = userRepository.getAll();
        User user;
        if (id==null){
            user = new User();
            user.setUserid(0);
            user.setColor(new Color(0,0,0));
            user.setPhoto(new AttachmentInfo(User.DEFAULT_PHOTO_FILENAME,
                                    User.DEFAULT_PHOTO_FILENAME));
        }else{
            user = userRepository.getById(id);
        }
        model.addAttribute("users", users);
        model.addAttribute("user", user);
        return "user/user_edit";
    }
    @PostMapping
    public String submitForm(@Valid User user, BindingResult result,
                @RequestPart("photo_control") Part photo,
                @RequestPart("attachments_control") Part[] attachments,
                Locale locale, Model model){
        //页面内即使不提交文件，也会为 file 控件生成非空 Part
        //此时 part.getSubmittedFileName()==""
        //因此参数 photo、attachments !=null

        if (result.hasErrors()){        //当未能通过验证
            List<User> users = userRepository.getAll();
            if (user.getUserid()!=0){
                User old = userRepository.getById(user.getUserid());
                if (old!=null) {
                    user.setCreatime(old.getCreatime());
                    user.setPhoto(old.getPhoto());
```

```java
                    user.setAttachments(old.getAttachments());
                }
            }else{
                user.setCreatime(LocalDateTime.now());
                user.setPhoto(new AttachmentInfo(User.DEFAULT_PHOTO_FILENAME,
                                    User.DEFAULT_PHOTO_FILENAME));
            }
            model.addAttribute("users", users);
            model.addAttribute("user", user);
            return "user/user_edit";
        }
        transferUploadedToUserInfo(user, photo, attachments);
        userRepository.save(user);
        return "redirect:/user?lang=" + locale;
    }
    @GetMapping("/download")
    public void download(@RequestParam("uid") Integer userid,
                 @RequestParam("filename") String fileTodownload,
                 HttpServletResponse resp) throws IOException {
        //查找附件的显示名称
        User user = userRepository.getById(userid);
        String fileNameDisplayed = "";
        for (int i=0;i<user.getAttachments().size();i++){
            if (user.getAttachments().get(i).getSavedFileName()
                                .equals(fileTodownload)){
                fileNameDisplayed = user.getAttachments().get(i)
                                .getUploadedFileName();
                break;
            }
        }
        File file = new File("D:/ImageOutside/" + fileTodownload);
        if (!file.exists()) {
            return ;
        }
        InputStream is = new FileInputStream(file);
        resp.setContentType("application/force-download");
        resp.addHeader("Content-Disposition", "attachment;fileName=" +
                            URLEncoder.encode(fileNameDisplayed, "UTF-8"));
        is.transferTo(resp.getOutputStream());
        is.close();
    }
    @GetMapping("/delete")
    public String delete(@RequestParam("uid") Integer userid,
                 @RequestParam("filename") String fileToDelete,
```

```java
                    Locale locale){
        //删除数据库的记录及外部文件
        userRepository.deleteAttachmentByFileName(userid, fileToDelete);
        return "redirect:/user?uid=" + userid + "&lang=" + locale;
    }
    @GetMapping("/remove")
    public String remove(@RequestParam("uid") Integer userid, Locale locale){
        userRepository.deleteById(userid);
        return "redirect:/user?lang=" + locale;
    }
    private void transferUploadedToUserInfo(User user, Part part, Part[] attachments){
        //界面上是新增用户
        if (user.getUserid()==0){
            user.setCreatime(LocalDateTime.now());
            //附件信息
            appendAttachmentsToUserInfo(user, attachments);
            //照片上传
            AttachmentInfo info = CommonUtil.saveUploadPart(part);
            if (info==null){         //无照片上传
                info = new AttachmentInfo(User.DEFAULT_PHOTO_FILENAME,
                                User.DEFAULT_PHOTO_FILENAME);
            }
            user.setPhoto(info);
        }else{                      //界面上是修改用户
            User oldUser = userRepository.getById(user.getUserid());
            user.setCreatime(oldUser.getCreatime());
            //表单不会提交原有附件信息，所以转换出来的 user 不含原有附件信息
            //从 Repository 读取此 user 的原有附件信息
            user.setAttachments(oldUser.getAttachments());
            //把表单提交的新附件信息添加到此 user 的原有附件信息之后
            appendAttachmentsToUserInfo(user, attachments);
            //照片上传
            AttachmentInfo info = CommonUtil.saveUploadPart(part);
            if (info==null){         //无照片上传
                info = oldUser.getPhoto();
            }
            user.setPhoto(info);
        }
    }
    private void appendAttachmentsToUserInfo(User user, Part[ ] parts){
        //若有附件上传，则新增的附件会追加至 user.attachments 末尾
        //若无附件上传，则不会更改 user.attachments
        for (Part part : parts){
            AttachmentInfo info =CommonUtil.saveUploadPart(part);
```

```
            if (info!=null){
                List<AttachmentInfo> attaches = user.getAttachments();
                if (attaches==null){
                    attaches = new ArrayList<>();
                }
                attaches.add(info);
                user.setAttachments(attaches);
            }
        }
    }
}
```

第八步：运行程序，之后在浏览器地址栏中输入 http://localhost:8081/user，查看显示结果。

12.5 用SpringBoot实现

设计思路

与 Spring 实现方案的设计思路相同。实现步骤如下：

第一步：在模块的 pom.xml 中添加 JPA 依赖。

代码 12-15：JPA 依赖

```
<dependency>
    <groupId>org.springframework.boot</groupId>
    <artifactId>spring-boot-starter-data-jpa</artifactId>
    <version>${springboot.version}</version>
</dependency>
<dependency>
    <groupId>mysql</groupId>
    <artifactId>mysql-connector-java</artifactId>
    <version>8.0.31</version>
</dependency>
```

第二步：新增 DatabaseConf 类，对数据库进行配置，与 12.4 节第二步相同。或者，删除 DatabaseConf 类，然后只需在 application.properties 内增加如下配置项：

- spring.datasource.driver-class-name=com.mysql.cj.jdbc.Driver
- spring.datasource.url=jdbc:mysql://localhost:3306/servletdb?serverTimezone=Asia/Shanghai&characterEncoding=utf8&useSSL=false
- spring.datasource.username=root
- spring.datasource.password=abcd1234
- spring.jpa.hibernate.ddl-auto=update
- spring.jpa.show-sql=false
- spring.data.jpa.repositories.enabled=true

SpringBoot 会根据这些配置项自动完成 JPA 相关对象的构建。

第三步：给 User 类增加 JPA 注解，与 12.4 节第三步相同。
第四步：新增 R_Attachments 类、R_Photo 类、R_Languages 类，与 12.4 节第四步相同。
第五步：新增 UserRepositoryDB 类，与 12.4 节第五步相同。
第六步：修改 user_edit.html，添加"删除"按钮用于删除界面内当前用户。与 12.4 节第六步相同。
第七步：更新 UserController 类。与 12.4 节第七步相同。
第八步：运行程序，之后在浏览器地址栏中输入 http://localhost:8082/user，查看显示结果。

12.6 小结

Persistence Context 可看作关系数据库表格记录的一个缓冲区，表格的部分或者全部记录在该缓冲区内以对象（称为持久化对象）的形式存在。EntityManager 对缓冲区进行管理，比如把普通对象放入缓冲区使之成为持久化对象，修改、删除缓冲区内的对象，当事务提交时，这些更改都会作用到表格记录上；EntityManager 对缓冲区的管理还包括把持久化对象从缓冲区移除使其成为普通对象，在缓冲区内查找对象（如果缓冲区内不存在此对象则自动从表格中查询记录并将其存入缓冲区）等。JPA 定义的操作接口都是围绕如何构建、维护及管理 Persistence Context 进行的。

12.7 习题

为本章案例的用户信息增加"民族"字段，在录入及编辑界面上此字段以列表框的形式让用户进行选择。要求信息保存到 MySQL 数据库以及支持国际化。

第 13 章
RESTful 服务

本章介绍如何使用 Servlet 提供 RESTful Web 服务。本章案例是通过浏览器或者 jersey 客户端访问 Servlet 所提供的 RESTful 服务。

通过学习本章内容，读者将可以：
- 使用 Servlet 提供 RESTful 服务

13.1 相关概念

HTTP 协议的应答消息头 Content-Type 指明了应答消息体的媒体类型。在 RESTful 服务中，Content-Type 通常设置为 application/json，这表示消息体采用 JSON 数据格式。RESTful 服务不仅支持 HTTP 的 GET 与 POST 请求，还支持 HTTP 的 PUT、DELETE、PATCH、HEAD 和 OPTIONS 请求。

13.1.1 RESTful

REST（REpresentational State Transfer，表征性状态转移）指的是一组架构约束条件和原则。符合下述 REST 主要原则的架构方式称为 RESTful：
- 对网络上所有资源都有一个资源标识符
- 对资源的操作不会改变标识符
- 同一资源有多种表现形式（如 xml、json 等）
- 所有操作都是无状态的

REST 本身并没有创造新的技术、组件或服务，而 RESTful 的理念就是使用 Web 的现有特征和能力，更好地使用现有 Web 标准中的一些准则和约束。目前 HTTP 是唯一与 REST 相关的实例，因此通常所说的 REST 就是通过 HTTP 实现的 REST。

在客户端和服务端之间传送的是资源的表述。资源的表述包括数据和描述数据的元数据，例如，HTTP 头部的 Content-Type 属性就是这样一个元数据属性。双方这样通过 HTTP 进行协商：客户端通过 HTTP 请求的头部 Accept 属性请求一种特定格式的表述，服务端则通过 HTTP 应答的头部 Content-Type 属性告诉客户端资源的表述形式。若服务器不支持所请求的表述格式，它应该返回一个 HTTP 406 状态码，表示拒绝处理该请求。

Jakarta-restful-ws-spec 是 RESTful Web 服务的 Java API 规范，是使用 Java 开发 REST 服务时的基本约定。

13.1.2 非阻塞输入

ServletInputStream 的 setReadListener(ReadListener)方法用于向其注册 ReadListener。注册 ReadListener 后将启动非阻塞输入，这时再使用阻塞式输入是不合法的。非阻塞输入意味着当 ServletInputStream 的数据可以被读取时，Servlet 容器会调用 ReadListener 的回调方法，程序必须在回调方法中读取输入流数据。ReadListener 提供以下回调方法供 Servlet 容器在对应的事件发生时调用：

- void onDataAvailable();
- void onAllDataRead();
- void onError(Throwable t)。

第一个函数的调用发生在 ServletInputStream 的流数据从不可读取变成可以读取的转折点（包括注册后第一次可以读取数据时），即 ServletInputStream 的 isReay()方法值从 false 向 true 转变的瞬间；第二个函数的调用发生在 ServletInputStream 的所有流数据都已经被读取时，此时 ServletInputStream 的 isFinished()方法返回 true；ServletInputStream 发生错误或者异常将导致第三个函数被 Servlet 容器调用。

13.2 案例描述

在浏览器地址栏中输入 http://localhost:8080/rs/user，则浏览器以 JSON 格式显示全部用户信息，显示效果如图 13-1 所示；在浏览器地址栏中输入 http://localhost:8080/rs/user/7，则浏览器以 JSON 格式显示 id 为 7 的用户信息，显示效果如图 13-2 所示。

图13-1 案例运行界面（1）

图13-2 案例运行界面（2）

说明：地址栏中的 7 是路径参数，可以替换成其他 id 值。

13.3 用Servlet实现

设计思路

新增若干非 Java 基础类（Gender、Major、PLanguage、Color）的序列/反序列化器，在 User 类对应的成员变量上用注解@JsonSerialize 及@JsonDeserialize 指明需要使用的序列化器及反序列化器。

新建 RsUserServlet，对于 GET 方法的请求，RsUserServlet 把从数据库查询到的 User 对象或者 User 对象列表转换成 JSON 格式再从 ServletResponse 的输出流输出；对于 POST 或者 PUT 方法的请求，RsUserServlet 以阻塞的方式读取 ServletRequest 的输入流或者外部文件的输入流，按照 JSON 格式将其转换成 User 对象或者 User 对象列表，之后从 ServletResponse 作出应答。

新建 AsyncRsUserServlet，对于 GET 方法的请求，直接移交（dispatch）至 RsUserServlet 处理；对于 POST 或者 PUT 方法，AsyncRsUserServlet 启动 ReadListener，ReadListener 以非阻塞的方式读取 ServletRequest 的输入流并将其保存至临时文件，当从输入流读取完毕时，ReadListener 将请求移交（dispatch）给 RsUserServlet 处理，而临时文件所保存的位置则以 ServletRequest 的命名属性作为参数传入。模块设计如图 13-3 所示。

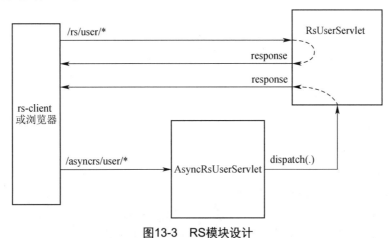

图13-3　RS模块设计

实现步骤如下：

第一步：在模块的 pom.xml 中添加依赖。

代码 13-1：相关的依赖

```xml
<dependency>
    <groupId>jakarta.ws.rs</groupId>
    <artifactId>jakarta.ws.rs-api</artifactId>
    <version>3.1.0</version>
</dependency>
<dependency>
    <groupId>com.fasterxml.jackson.core</groupId>
    <artifactId>jackson-databind</artifactId>
    <version>${jackson.version}</version>
</dependency>
<dependency>
    <groupId>com.fasterxml.jackson.datatype</groupId>
    <artifactId>jackson-datatype-jsr310</artifactId>
    <version>${jackson.version}</version>
</dependency>
```

第二步：新增若干非 Java 基础类的序列/反序列化器。

代码 13-2：EnumSerializer<T>类

```java
public class EnumSerializer<T> extends JsonSerializer<T> {
    private Class<T> clazz;
    public EnumSerializer(Class<T> clazz){
```

```
        this.clazz = clazz;
    }
    @Override
    public void serialize(T anEnum, JsonGenerator jsonGenerator,
                    SerializerProvider serializerProvider)
                         throws IOException {
        EnumConversion<T> enumToString = new EnumConversion<>();
        String str = enumToString.convertEnumToIndexString(anEnum, clazz);
        jsonGenerator.writeString(str);
    }
}
```

再基于 EnumSerializer<T>创建：

- GenderSerializer 类
- MajorSerializer 类
- PLanguageSerializer 类

代码 13-3：EnumDeserializer<T>类

```
public class EnumDeserializer<T> extends JsonDeserializer<T> {
    private Class<T> clazz;
    public EnumDeserializer(Class<T> clazz){
        this.clazz = clazz;
    }
    @Override
    public T deserialize(JsonParser jsonParser,
            DeserializationContext deserializationContext)
                throws IOException, JacksonException {
        String value =jsonParser.getText();
        Integer index = Integer.valueOf(value);
        EnumConversion<T> toEnum = new EnumConversion<>();
        return toEnum.convertIndexToEnum(index, clazz);
    }
}
```

再基于 EnumDeserializer<T>创建：

- GenderDeserializer 类
- MajorDeserializer 类
- PLanguageDeserializer 类

代码 13-4：ColorSerializer 类

```
public class ColorSerializer extends JsonSerializer<Color> {
    @Override
    public void serialize(Color color, JsonGenerator jsonGenerator,
                    SerializerProvider serializerProvider)
                         throws IOException {
```

```
            String valueColor = "";
            if(color!=null) {
                valueColor = CommonUtil.colorToHexString(color);
            }
            jsonGenerator.writeString(valueColor);
        }
    }
```

代码 13-5：ColorDeserializer 类

```
public class ColorDeserializer extends JsonDeserializer<Color> {
    @Override
    public Color deserialize(JsonParser jsonParser,
                DeserializationContext deserializationContext)
                    throws IOException, JacksonException {
        String value =jsonParser.getText();
        if (value==null || value.isBlank()){
            return null;
        }
        return Color.decode("0x" +value.substring(1));
    }
}
```

第三步：修改 User 实体类，为其中非 Java 基础类的成员变量增加@JsonSerialze 及 @JsonDeserialize 注解。

代码 13-6：增加序列化/反序列化注解

```
@JsonSerialize(using = GenderSerializer.class)
@JsonDeserialize(using = GenderDeserializer.class)
private Gender sex;

@JsonSerialize(using = ColorSerializer.class)
@JsonDeserialize(using = ColorDeserializer.class)
private Color color;

@JsonSerialize(contentUsing = PLanguageSerializer.class)
@JsonDeserialize(contentUsing = PLanguageDeserializer.class)
private List<PLanguage> languages;

@JsonSerialize(using = MajorSerializer.class)
@JsonDeserialize(using = MajorDeserializer.class)
private Major speciality;

@JsonSerialize(using = ColorSerializer.class)
@JsonDeserialize(using = ColorDeserializer.class)
private Color color;
```

```java
@JsonIgnore      //否则生成的json会包含一个colorString字段
public String getColorString(){
    return CommonUtil.colorToHexString(color);
}
```

第四步：修改CharacterEncodingFilter类，设置ContentType。

代码13-7：CharacterEncodingFilter类

```java
public class CharacterEncodingFilter implements Filter {
    @Override
    public void doFilter(ServletRequest servletRequest,
                ServletResponse servletResponse,
                FilterChain filterChain) throws IOException, ServletException {
        //用于解决form提交中文时服务端得到乱码的问题
        servletRequest.setCharacterEncoding("utf-8");
        //解决网页显示中文时乱码问题
        servletResponse.setCharacterEncoding("utf-8");

        HttpServletRequest request = (HttpServletRequest) servletRequest;
        //设置rs应答的json格式
        if (request.getServletPath().startsWith("/rs/")){
            servletResponse.setContentType("application/json");
        }
        filterChain.doFilter(servletRequest, servletResponse);
    }
}
```

第五步：新建监听器ReadListenerForRequest，将用于AsyncRsUserServlet的异步输入。

代码13-8：ReadListenerForRequest类

```java
public class ReadListenerForRequest implements ReadListener {
    public final static String REQUEST_TO_TEMP_FILE_ATTR_NAME =
            "cxiao.sh.cn.listener.ReadListenerForRequest.savedFilePath";
    private AsyncContext asct;
    private OutputStream os;
    private String savedFilePath = "D:/ImageOutside/" + UUID.randomUUID().toString();
    private String targetUrl;
    public ReadListenerForRequest(HttpServletRequest req,
                        HttpServletResponse resp,
                        String targetUrl) throws IOException {
        asct = req.startAsync(req, resp);
        asct.setTimeout(-1);
        this.targetUrl = targetUrl;
        File fileToSave = new File(savedFilePath);
        os = new FileOutputStream(fileToSave);
    }
```

```
    public void setup() throws IOException {
        asct.getRequest().getInputStream().setReadListener(this);
    }
    @Override
    public void onDataAvailable() throws IOException {
        int nRead;
        byte[ ] buffer = new byte[1024*1024];
        ServletInputStream is = asct.getRequest().getInputStream();
        while(is.isReady()){
            if((nRead = is.read(buffer, 0, buffer.length))>0){
                os.write(buffer, 0, nRead);
            }
        }
    }
    @Override
    public void onAllDataRead() throws IOException {
        os.close();
        asct.getRequest().setAttribute(REQUEST_TO_TEMP_FILE_ATTR_NAME,savedFilePath);
        asct.dispatch(this.targetUrl);
    }
    @Override
    public void onError(Throwable throwable) {
        asct.complete();
        try{
            os.close();
        }catch (Exception e){
            e.printStackTrace();
        }
    }
}
```

此类的作用：在监听器中把 HttpRequest 的输入信息保存至临时文件，当 HttpRequest 输入完毕时，移交至参数 targetUrl 对应的 Servlet 处理。通过 targetUrl 把 Servlet 变成回调目标，并通过 req.Attributes 传递如下参数：所生成的临时文件路径名。

第六步：新增 RsUserServlet 类。

代码 13-9：RsUserServlet 类

```
public class RsUserServlet extends HttpServlet {
    @Override
    protected void doGet(HttpServletRequest req, HttpServletResponse resp)
                        throws ServletException, IOException {
        IRepository<User, Integer> userRepository =
                RepositoryUtil.userRepository(req.getServletContext());
        ObjectMapper objectMapper = new ObjectMapper();
```

```java
            //用代码加载jackson-datatype-jsr310模块,用以支持 LocalDate 等类型
            objectMapper.findAndRegisterModules();
            //或者通过如下方法进行注册
            //objectMapper.registerModule(new JavaTimeModule());
            String path = req.getPathInfo();
            if (path==null){  //当用户访问/rs/user时,返回User列表
                List<User> users = userRepository.getAll();
                objectMapper.writeValue(resp.getOutputStream(), users);
                System.out.println("输出 JSON 至客户端: " + users);
            }else{             //当用户访问 /rs/user/419 时,返回 id 为 419 的 User 对象
                Integer uid = Integer.valueOf(path.substring(path.lastIndexOf("/") + 1));
                User user = userRepository.getById(uid);
                objectMapper.writeValue(resp.getOutputStream(), user);
                System.out.println("输出 JSON 至客户端: " + user);
            }
        }
    //PUT 请求用于接收 List<User>
    @Override
    protected void doPut(HttpServletRequest req, HttpServletResponse resp)
                    throws ServletException, IOException {
        IRepository<User, Integer> userRepository =
                    RepositoryUtil.userRepository(req.getServletContext());
        ObjectMapper objectMapper = new ObjectMapper();
        //用代码加载jackson-datatype-jsr31模块,用以支持 LocalDate 等类型
        //objectMapper.findAndRegisterModules();
        //或者通过如下方法进行注册
        objectMapper.registerModule(new JavaTimeModule());
        //设置解析 Json 的目标类型
        JavaType javaType = objectMapper.getTypeFactory()
                    .constructParametricType(ArrayList.class, User.class);
        List<User> users;
        //如果是从 AsyncRsUserServlet 移交来的
        if (req.getDispatcherType().equals(DispatcherType.ASYNC)){
            String savedFilePath = (String)req.getAttribute(
                    ReadListenerForRequest.REQUEST_TO_TEMP_FILE_ATTR_NAME);
            System.out.println("从 " + savedFilePath + " 读取 Put 信息! ");
            users = objectMapper.readValue(new File(savedFilePath), javaType);
            if (new File(savedFilePath).delete()){
                System.out.println("删除 Put 输入流的临时文件: " + savedFilePath);
            }
        }else {     //如果是从客户端直接提交的请求
            users = objectMapper.readValue(req.getInputStream(), javaType);
        }
```

```
            System.out.println("解析 PUT 得到的 List<User>: " + users);

            for(User user: users){
                userRepository.save(user);
            }
            resp.setStatus(HttpServletResponse.SC_OK);
        }
        //POST 请求用于接收 User
        @Override
        protected void doPost(HttpServletRequest req, HttpServletResponse resp)
                        throws ServletException, IOException {
            IRepository<User, Integer> userRepository =
                    RepositoryUtil.userRepository(req.getServletContext());
            ObjectMapper objectMapper = new ObjectMapper();
            //用代码加载 jackson-datatype-jsr310 模块，用以支持 LocalDate 等类型
            //objectMapper.findAndRegisterModules();
            //或者通过如下方法进行注册
            objectMapper.registerModule(new JavaTimeModule());
            User user;
            //如果是从 AsyncRsUserServlet 移交来的
            if (req.getDispatcherType().equals(DispatcherType.ASYNC)){
                String savedFilePath = (String)req.getAttribute(
                    ReadListenerForRequest.REQUEST_TO_TEMP_FILE_ATTR_NAME);
                System.out.println("从 " + savedFilePath + " 读取 Post 信息！");
                user = objectMapper.readValue(new File(savedFilePath), User.class);
                if (new File(savedFilePath).delete()){
                    System.out.println("删除 Post 输入流的临时文件: "+savedFilePath);
                }
            }else {          //如果是从客户端直接提交的请求
                user = objectMapper.readValue(req.getInputStream(), User.class);
            }
            System.out.println("解析 Post 得到的 User: " + user);
            userRepository.save(user);
            resp.setStatus(HttpServletResponse.SC_OK);
        }
    }
```

doGet(.)方法也可以从 objectMapper.writeValueAsBytes(.)的结果构造输入流然后采用异步输出。

第七步：新增 AsyncRsUserServlet 类。

代码 13-10：AsyncRsUserServlet 类

```
public class AsyncRsUserServlet extends HttpServlet {
    @Override
    protected void doGet(HttpServletRequest req, HttpServletResponse resp)
```

```java
                   throws ServletException, IOException {
    AsyncContext asct = req.startAsync(req, resp);
    asct.setTimeout(-1);
    if (req.getPathInfo()!=null) {
        asct.dispatch("/rs/user" + req.getPathInfo());
    }else{
        asct.dispatch("/rs/user");
    }
}
@Override
protected void doPost(HttpServletRequest req, HttpServletResponse resp)
                   throws ServletException, IOException {
    new ReadListenerForRequest(req, resp, "/rs/user").setup();
}
@Override
protected void doPut(HttpServletRequest req, HttpServletResponse resp)
                   throws ServletException, IOException {
    new ReadListenerForRequest(req, resp, "/rs/user").setup();
}
}
```

在 AsyncRsUserServlet 类中异步接收完用户提交的 JSON 请求后，再移交至 RsUserServlet 处理。

第八步：修改 ServletInitializer 类，增加对 RsUserServlet 及 AsyncRsUserServlet 的注册。

代码 13-11：注册 RsUserServlet 及 AsyncRsUserServlet

```java
//注册 RsUserServlet
servlet = new RsUserServlet();
registration = servletContext.addServlet("rs.user.servlet", servlet);
registration.setLoadOnStartup(6);
registration.addMapping("/rs/user/*");
registration.setAsyncSupported(true);

//注册 AsyncRsUserServlet
servlet = new AsyncRsUserServlet();
registration = servletContext.addServlet("asynrs.user.servlet", servlet);
registration.setLoadOnStartup(7);
registration.addMapping("/asyncrs/user/*");
registration.setAsyncSupported(true);
```

第九步：运行程序，之后在浏览器地址栏里输入如下网址：

http://localhost:8080/rs/user

http://localhost:8080/asyncrs/user

http://localhost:8080/rs/user/14

http://localhost:8080/asyncrs/user/14

查看显示结果。并运行 rs-client 模块测试 Rs 服务。

13.4 用jersey框架实现rs-client

设计思路

新建基于 jersey 框架的 rs-client 模块，可以发出 HTTP 的 GET、POST、PUT 请求。用于测试 RESTful 服务。实现步骤如下：

第一步：添加 jersey 依赖。

代码 13-12：jersey 依赖

```xml
<dependency>
    <groupId>org.projectlombok</groupId>
    <artifactId>lombok</artifactId>
    <version>1.18.24</version>
</dependency>
<!-- Jakart-RS 3.x，内含客户端API-->
<dependency>
    <groupId>jakarta.ws.rs</groupId>
    <artifactId>jakarta.ws.rs-api</artifactId>
    <version>3.1.0</version>
</dependency>
<!-- 有关Jersey的依赖-->
<dependency>
    <groupId>org.glassfish.jersey.core</groupId>
    <artifactId>jersey-server</artifactId>
    <version>${jersey.version}</version>
</dependency>
<dependency>
    <groupId>org.glassfish.jersey.containers</groupId>
    <artifactId>jersey-container-servlet</artifactId>
    <version>${jersey.version}</version>
</dependency>
<dependency>
    <groupId>org.glassfish.jersey.containers</groupId>
    <artifactId>jersey-container-servlet-core</artifactId>
    <version>${jersey.version}</version>
</dependency>
<dependency>
    <groupId>org.glassfish.jersey.inject</groupId>
    <artifactId>jersey-hk2</artifactId>
    <version>${jersey.version}</version>
</dependency>
```

```xml
<dependency>
    <groupId>org.glassfish.jersey.media</groupId>
    <artifactId>jersey-media-json-jackson</artifactId>
    <version>${jersey.version}</version>
</dependency>
<dependency>
    <groupId>com.fasterxml.jackson.core</groupId>
    <artifactId>jackson-databind</artifactId>
    <version>2.14.2</version>
</dependency>
<dependency>
    <groupId>com.fasterxml.jackson.datatype</groupId>
    <artifactId>jackson-datatype-jsr310</artifactId>
    <version>2.14.2</version>
</dependency>
<dependency>
    <groupId>org.glassfish.jersey.media</groupId>
    <artifactId>jersey-media-moxy</artifactId>
    <version>${jersey.version}</version>
</dependency>
<dependency>
    <groupId>org.glassfish.jersey.media</groupId>
    <artifactId>jersey-media-jaxb</artifactId>
    <version>${jersey.version}</version>
</dependency>
<dependency>
    <groupId>org.glassfish.jersey.media</groupId>
    <artifactId>jersey-media-multipart</artifactId>
    <version>${jersey.version}</version>
</dependency>
<dependency>
    <groupId>org.glassfish.jersey.media</groupId>
    <artifactId>jersey-media-sse</artifactId>
    <version>${jersey.version}</version>
</dependency>
<dependency>
    <groupId>org.glassfish.jersey.core</groupId>
    <artifactId>jersey-client</artifactId>
    <version>${jersey.version}</version>
</dependency>
```

第二步：新增以下若干枚举类，类的定义与 13.3 节 Servlet 方案所用类相同。

❑ Gender

❑ Major

- PLanguage
- EnumConversion

第三步：新增以下若干序列化/反序列化类，类的定义与 13.3 节 Servlet 方案所用类相同。

- ColorSerializer
- ColorDeserializer
- EnumSerializer
- EnumDeserializer
- GenderSerializer
- GenderDeserializer
- MajorSerializer
- MajorDeserializer
- PLanguageSerializer
- PLanguageDeserializer

第四步：新增 CommonUtil 类，类的定义与 13.3 节 Servlet 方案所用类相同，并增加 getRandomString(.)函数。此函数的功能是产生指定长度的随机字符串。

代码 13-13：getRandomString(.)

```
public static String getRandomString(int length){
    String str="abcdefghijklmnopqrstuvwxyzABCDEFGHIJKLMNOPQRSTUVWXYZ";
    Random random=new Random();
    StringBuffer sb=new StringBuffer();
    for(int i=0;i<length;i++){
        int number=random.nextInt(str.length());
        sb.append(str.charAt(number));
    }
    return sb.toString();
}
```

第五步：新增以下实体类。

- AttachmentInfo
- User

AttachmentInfo 类的定义与 13.3 节 Servlet 方案所用类相同。User 类的定义如下：

代码 13-14：User 类

```
@Data
@NoArgsConstructor
@AllArgsConstructor
public class User {
    private Integer userid;
    private String account;
    private String password;
    private String name;
```

```
@JsonSerialize(using = GenderSerializer.class)
@JsonDeserialize(using = GenderDeserializer.class)
private Gender sex;
private double height;
private Integer weight;
private LocalDate birthday;
@JsonSerialize(using = ColorSerializer.class)
@JsonDeserialize(using = ColorDeserializer.class)
private Color color;
@JsonIgnore
public String getColorString(){
    return CommonUtil.colorToHexString(color);
}
@JsonSerialize(contentUsing = PLanguageSerializer.class)
@JsonDeserialize(contentUsing = PLanguageDeserializer.class)
private List<PLanguage> languages;
@JsonSerialize(using = MajorSerializer.class)
@JsonDeserialize(using = MajorDeserializer.class)
private Major speciality;
private String mail;
private String address;
private AttachmentInfo photo;
private List<AttachmentInfo> attachments;
private LocalDateTime creatime;
}
```

第六步：新增 Client 类。

代码 13-15：Client 类

```
public class Client {
    private WebTarget target;
    private MediaType mediaType = MediaType.APPLICATION_JSON_TYPE;
    public Client(){
        ClientConfig clientConfig = new ClientConfig();
        //用于支持 XML 格式
        clientConfig.register(MoxyXmlFeature.class);
        //用于支持 JSON 格式
        clientConfig.register(JacksonFeature.class);
        //用于支持上传附件
        clientConfig.register(MultiPartFeature.class);

        jakarta.ws.rs.client.Client rsClient = ClientBuilder.newClient(clientConfig);
        target = rsClient.target("http://localhost:8080/asyncrs");
        //测试时可换成 /rs
    }
```

```java
public static void main(String[] args) {
    Client client = new Client();
    client.doWork();
}
public void doWork(){
    getAllUsers();
    getUserById(7);
    postUser();
    putUsers(5);
}
private User getUserById(Integer id){
    Response response = target
            .path("user")
            .path(String.valueOf(id))
            .request(mediaType)
            .get();
    User user = response.readEntity(User.class);
    System.out.println("getUserById(): \n" + user);
    return user;
}
private List<User> getAllUsers(){
    Response response = target
            .path("user")
            .request(mediaType)
            .get();
    GenericType<List<User>> genericType = new GenericType<>(){};
    List<User> users = response.readEntity(genericType);
    System.out.println("getAllUsers(): \n" + users);
    return users;
}
//构造一个随机User对象
private User randomUser(){
    List<User> users = getAllUsers();
    User user = users.get(new Random().nextInt(users.size()));
    user.setAccount(CommonUtil.getRandomString(8));
    user.setName(CommonUtil.getRandomString(15));
    user.setUserid(0);
    return user;
}
//构造n个随机User对象
private List<User> multipleUsers(int n){
    List<User> users = new ArrayList<>();
    List<User> users_db = getAllUsers();
    for (int i=0;i<n;i++){
```

```java
            User user = users_db.get(new Random().nextInt(users_db.size()));
            user.setAccount(user.getAccount() + "9");
            user.setName(user.getName() + "8");
            //随机选择是新增用户还是修改用户
            if(new Random().nextInt(2)==0){
                user.setUserid(0);
            }
            users.add(user);
        }
        return users;
    }
    private void postUser(){
        User user = randomUser();
        Entity<User> userEntity = Entity.entity(user, mediaType);
        Response response = target
                .path("user")
                .request(mediaType)
                .post(userEntity);

        if (response.getStatus()==Response.Status.OK.getStatusCode()){
            System.out.println("Post 成功! ");
        }else{
            System.out.println("Post 失败! ");
        }
    }
    private void putUsers(int n){
        List<User> users = multipleUsers(n);
        Entity<List<User>> entity = Entity.entity(users, mediaType);
        Response response = target
                .path("user")
                .request(mediaType)
                .put(entity);
        if (response.getStatus()==Response.Status.OK.getStatusCode()){
            System.out.println("Put 成功! ");
        }else{
            System.out.println("Put 失败! ");
        }
    }
}
```

为了便于测试，需要手动把 MySQL 数据库中表格 r_attachments 的主键设为（uid, savedfilename）联合主键，同时给 Servlet 模块中 R_Attachments 的 uid 属性添加@Id 注解，原有 savedfilename 属性上的@Id 注解不变。

如果 rs-client 对 Spring 模块测试，则把 rs-client 访问的服务端口号更改为 8081；如果 rs-client

对 SpringBoot 模块测试，则把 rs-client 访问的服务端口号更改为 8082。

13.5 用Spring实现

设计思路

添加 JSON 格式文本与对象的转换器 MappingJackson2HttpMessageConverter 对象，它可以完成 Java 基础类对象的转换；对于非 Java 基础类，则自定义序列/反序列化器。在 Controller 类的方法参数上标注@RequestBody，则该参数对象来自 HTTP 请求消息体的转换；在 Controller 类上标注@RestController，则此 Controller 类的方法返回值将转换成 HTTP 应答消息体发送至客户端。实现步骤如下：

第一步：添加 JSON 依赖。

代码 13-16：JSON 依赖

```
<dependency>
    <groupId>com.fasterxml.jackson.core</groupId>
    <artifactId>jackson-databind</artifactId>
    <version>${jackson.version}</version>
</dependency>
<dependency>
    <groupId>com.fasterxml.jackson.datatype</groupId>
    <artifactId>jackson-datatype-jsr310</artifactId>
    <version>${jackson.version}</version>
</dependency>
```

第二步：新增非 Java 基础类的序列/反序列化器，与 13.3 节第二步相同。

第三步：修改 User 实体类，为其中非 Java 基础类的成员变量增加@JsonSerialze 及@JsonDeserialize 注解，与 13.3 节第三步相同。

第四步：在 MvcConfig 类的 configureMessageConverters(.)方法内添加转换器。

代码 13-17：添加转换器

```
//把对象转成 JSON 格式的转换器
ObjectMapper objectMapper = new ObjectMapper();
objectMapper.enable(SerializationFeature.INDENT_OUTPUT);
objectMapper.setDateFormat(new SimpleDateFormat("yyyy年MM月dd日"));
//用于解决 LocalDate、LocalDateTime 类型转换
objectMapper.registerModule(new JavaTimeModule());
objectMapper.setSerializationInclusion(JsonInclude.Include.NON_EMPTY);
MappingJackson2HttpMessageConverter converter = new
                    MappingJackson2HttpMessageConverter(objectMapper);
converter.setPrettyPrint(true);
converter.setDefaultCharset(StandardCharsets.UTF_8);
converters.add(converter);
```

第五步：新增 RsUserController 类。

代码 13-18：RsUserController 类

```java
@RestController
@RequestMapping(path="/rs/user")
public class RsUserController {
    @Autowired
    IRepository<User, Integer> userRepository;
    @GetMapping
    public List<User> listUsers(){
        List<User> users = userRepository.getAll();
        return users;
    }
    @GetMapping(path="/{uid}")
    public User queryUserById(@PathVariable("uid") Integer userid){
        System.out.println("客户查询的 user id = " + userid);
        User user = userRepository.getById(userid);
        System.out.println(user);
        return user;
    }
    @PostMapping
    public void saveUser(@RequestBody User user){
        System.out.println("客户 Post 的 User:" + user);
        userRepository.save(user);
    }
    @PutMapping
    public void saveUsers(@RequestBody List<User> users){
        System.out.println("客户 PUT 的 List<User>:" + users);
        for (User user: users){
            userRepository.save(user);
        }
    }
}
```

第六步：运行程序，之后在浏览器地址栏中输入 http://localhost:8081/rs/user 或者 http://localhost:8081/rs/user/14，查看显示结果。并运行 rs-client 模块测试 Rs 服务。

13.6 用SpringBoot实现

设计思路

与 Spring 实现方案的设计思路相同。实现步骤如下：

第一步：添加 JSON 依赖。

代码 13-19：JSON 依赖

```xml
<dependency>
    <groupId>org.springframework.boot</groupId>
    <artifactId>spring-boot-starter-json</artifactId>
    <version>${springboot.version}</version>
</dependency>
```

第二步：新增非 Java 基础类的序列/反序列化器，与 13.5 节第二步相同。

第三步：修改 User 实体类，为其中非 Java 基础类的成员变量增加@JsonSerialze 及 @JsonDeserialize 注解，与 13.5 节第三步相同。

第四步：新增 RsUserController 类，与 13.5 节第五步相同。

第五步：运行程序，之后在浏览器地址栏中输入 http://localhost:8082/rs/user 或者 http://localhost:8082/rs/user/14，查看显示结果。并运行 rs-client 模块测试 Rs 服务。

13.7 小结

当 Servlet 处理浏览器提交的网页表单时，Servlet 需要进行输入数据绑定与输出数据绑定。类似地，当 Servlet 提供 RESTful 服务时，Servlet 也需要进行输入数据"绑定"与输出数据"绑定"，输入数据"绑定"是把 JSON 格式文本转换成对象，而输出数据"绑定"是把对象转换成 JSON 格式文本，这种双向转换可以由 ObjectMapper 和自定义的序列/反序列化器完成。应答消息体的 JSON 格式使得应答消息头 Content-Type 必须设置为 application/json。提供 RESTful 服务的 Servlet 也可以使用异步处理，如移交请求、非阻塞式输入/输出。

13.8 习题

为本章案例的 RESTful 服务增加 DELETE 请求处理：当 rs-client 以 HTTP 的 DELETE 方法访问 http://localhost:8080/rs/user/7 时，Servlet 会删除数据库内 id 为 7 的用户信息并向 rs-client 返回 OK 状态码；如果数据库中不存在 id 为 7 的用户信息或者删除操作出错，则返回"服务器错误"状态码。上述 URL 中数字 7 是路径参数，使用时可以替换成其他整数。

第 14 章
SSE

本章介绍如何使用 Servlet 提供事件推送服务。本章案例是编写一个模拟落球的应用。
通过学习本章内容，读者将可以：
❑ 用 Servlet 提供服务端的推送服务

14.1 相关概念

服务器推送事件（Server-Sent Events，SSE）是一种基于 HTTP 的、以流的形式由服务端持续向客户端发送数据的技术。SSE 具有"一次请求，多次应答"的特点。

14.1.1 SSE特点

HTTP 协议采用"请求-应答"模式，当使用非 KeepAlive 模式时，客户端和服务端针对每个请求-应答新建一个连接，完成之后立即断开连接；当使用 KeepAlive 模式时，KeepAlive 功能使客户端到服务端的连接持续有效。JavaScript 的 EventSource 对象可以与 Servlet 建立 KeepAlive 模式的连接，这种持久连接允许 Servlet 对客户端的一次请求进行连续的多次应答，当应答消息头 Content-Type 设置为 text/event-stream 时，这些应答就是服务端推送至客户端的事件。事件是轻量级文本，有如下四个属性可供设置：

❑ id：事件的 ID
❑ event：事件的类型
❑ data：事件的数据
❑ retry：事件流的重连接时间

客户端无法在事件通道上作出二次请求或者对服务端推送的事件再作出"响应"，即无法在同一连接上支持复杂的交互需求。

14.1.2 事件队列

事件源与 Servlet 之间通过线程安全的队列传递事件。事件源是队列元素的生产者，Servlet 是队列元素的消费者。线程安全队列可以分为阻塞队列和非阻塞队列，在并发编程中，推荐使用阻塞队列，这样可以尽量地避免程序出现意外的错误；如果使用非阻塞队列，虽然能即时返回结果，但必须自行编码处理返回为空的情况以及消费重试等问题。

Java 中接口 BlockingQueue 描述了阻塞队列，而接口 ConcurrentLinkedQueue 描述了非阻塞队列。ArrayBlockingQueue 是基于数组的阻塞队列实现，而 LinkedBlockingQueue 是基于链表的阻塞队列实现。

表 14-1 列出了接口 BlockingQueue 的常用方法。

表 14-1 接口 BlockingQueue 提供的常用方法

	可能抛出异常	返回布尔值	可 能 阻 塞	设定等待时间
入队	add(e)	offer(e)	put(e)	offer(e, timeout, unit)
出队	remove()	poll()	take()	poll(timeout, unit)
查看	element()	peek()	无	无

说明如下：

（1）方法 add(e)、remove()、element()不会阻塞线程，当不满足约束条件时，会抛出 IllegalStateException 异常；例如，当队列被元素填满后，再调用 add(e)，则会抛出异常。

（2）方法 offer(e)、poll()、peek()不会阻塞线程，也不会抛出异常；例如，当队列被元素填满后，再调用 offer(e)，则不会插入元素，函数返回 false。

（3）方法 put(e)、take()当不满足约束条件时，会阻塞线程；例如，当队列被元素填满后，再调用 put(e)，则阻塞当前线程。

14.2 案例描述

系统运行效果如图 14-1 所示，不同形状的黑色球体依次从顶部落下，进入底部的框内。

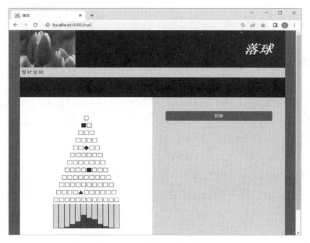

图14-1 案例运行界面

14.3 用Servlet实现

设计思路

新增两个 Servlet，一个是 BallServlet，此 Servlet 用于处理用户界面操作；另一个是 SseServlet，此 Servlet 接收 JavaScript 发送的请求，启动 WriteListener，以非阻塞的方式向 JavaScript 推送事件流（即源源不断地从 ServletResponse 输出），SseServlet 的模块设计如图 14-2 所示。服务端事件源在一个会话的首次请求时被创建并以命名属性保存于 HttpSession 中。

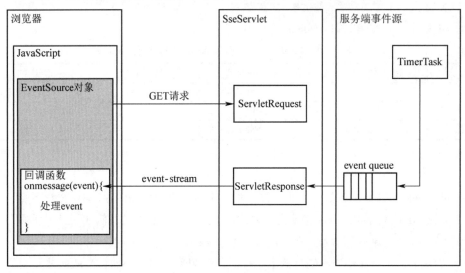

图14-2　SseServlet模块流程

实现步骤如下：

第一步：新增如下自定义的 SSE 基础类。

- EventSource
- OutboundEvent
- SSEWriteListener
- SseServlet

代码 14-1：EventSource 类

```java
public abstract class EventSource extends TimerTask {
    private BlockingQueue<String> data = new ArrayBlockingQueue<>(10);
    private final Semaphore runFlag = new Semaphore(1, true);
    protected Timer timer = new Timer();
    abstract protected void initialize();
    abstract protected String produceOneEvent();
    public String getEvent() throws InterruptedException{
        return this.getDataQueue().take();
    }
    protected BlockingQueue<String> getDataQueue(){
```

```java
            return this.data;
        }
        @Override
        public void run() {
            try{
                runFlag.acquire();
            }catch (InterruptedException e){
                e.printStackTrace();
            }
            String event = produceOneEvent();
            if (event!=null) {        //等于null表示没有Event产生
                try{
                    this.getDataQueue().put(event);
                }catch (InterruptedException e){
                    e.printStackTrace();
                }
            }
            runFlag.release();
        }
        public void pause(){
            try {
                runFlag.acquire();
            }catch (InterruptedException ex){
                ex.printStackTrace();
            }
        }
        public void resume(){
            runFlag.release();
        }
    }
```

代码 14-2：OutboundEvent 类

```java
public final class OutboundEvent implements OutboundSseEvent {
    private final String name;
    private final String comment;
    private final String id;
    private final GenericType type;
    private final MediaType mediaType;
    private final Object data;
    private final long reconnectDelay;
    OutboundEvent(String name, String id, long reconnectDelay,
            GenericType type, MediaType mediaType,
```

```java
                  Object data, String comment) {
    this.name = name;
    this.comment = comment;
    this.id = id;
    this.reconnectDelay = reconnectDelay;
    this.type = type;
    this.mediaType = mediaType;
    this.data = data;
}
@Override
public String toString() {
    String s = "";
    s += "id:" + getId() + "\n";
    s += "data:" + getData().toString() + "\n";
    s += "name:" + getName() + "\n";
    s += "comment:" + getComment() + "\n";
    s += "reconnectDelay:" + getReconnectDelay() + "\n";
    s += "\n";
    return s;
}
public String getName() {
    return this.name;
}
public String getId() {
    return this.id;
}
public long getReconnectDelay() {
    return this.reconnectDelay;
}
public boolean isReconnectDelaySet() {
    return this.reconnectDelay > -1L;
}
public Class<?> getType() {
    return this.type == null ? null : this.type.getRawType();
}
public Type getGenericType() {
    return this.type == null ? null : this.type.getType();
}
public MediaType getMediaType() {
    return this.mediaType;
}
```

```java
    public String getComment() {
        return this.comment;
    }
    public Object getData() {
        return this.data;
    }
    public static class Builder implements OutboundSseEvent.Builder {
        private String name;
        private String comment;
        private String id;
        private long reconnectDelay = -1L;
        private GenericType type;
        private Object data;
        private MediaType mediaType;

        public Builder() {
            this.mediaType = MediaType.TEXT_PLAIN_TYPE;
        }

        public Builder name(String name) {
            this.name = name;
            return this;
        }
        public Builder id(String id) {
            this.id = id;
            return this;
        }
        public Builder reconnectDelay(long milliseconds) {
            if (milliseconds < 0L) {
                milliseconds = -1L;
            }
            this.reconnectDelay = milliseconds;
            return this;
        }
        public Builder mediaType(MediaType mediaType) {
            if (mediaType == null) {
                throw new NullPointerException("out.event.media.type.null");
            } else {
                this.mediaType = mediaType;
                return this;
            }
```

```java
            }
            public Builder comment(String comment) {
                this.comment = comment;
                return this;
            }
            public Builder data(Class type, Object data) {
                if (data == null) {
                    throw new NullPointerException("out.event.data.null");
                } else if (type == null) {
                    throw new NullPointerException("out.event.data.type.null");
                } else {
                    this.type = new GenericType(type);
                    this.data = data;
                    return this;
                }
            }
            public Builder data(GenericType type, Object data) {
                if (data == null) {
                    throw new NullPointerException("out.event.data.null");
                } else if (type == null) {
                    throw new NullPointerException("out.event.data.type.null");
                } else {
                    this.type = type;
                    if (data instanceof GenericEntity) {
                        this.data = ((GenericEntity)data).getEntity();
                    } else {
                        this.data = data;
                    }
                    return this;
                }
            }
            public Builder data(Object data) {
                if (data == null) {
                    throw new NullPointerException("out.event.data.null");
                } else {
                    return this.data(genericTypeFor(data), data);
                }
            }
            private static GenericType genericTypeFor(Object instance) {
                GenericType genericType;
                if (instance instanceof GenericEntity) {
```

```java
                    genericType = new GenericType(((GenericEntity)instance).getType());
            } else {
                genericType = instance ==
                        null ? null : new GenericType(instance.getClass());
            }
            return genericType;
        }
        public OutboundEvent build() {
            if(this.comment == null && this.data == null && this.type == null) {
                throw new IllegalStateException("out.event.not.buildable");
            } else {
                return new OutboundEvent(this.name, this.id, this.reconnectDelay,
                        this.type, this.mediaType, this.data, this.comment);
            }
        }
    }
}
```

代码 14-3：SSEWriteListener 类

```java
public class SSEWriteListener implements WriteListener {
    private AsyncContext asct;
    private EventSource eventSource;
    public SSEWriteListener(EventSource eventSource,
                HttpServletRequest request,
                HttpServletResponse response){
        this.eventSource = eventSource;
        asct = request.startAsync(request, response);
        asct.getResponse().setContentType("text/event-stream");
        asct.setTimeout(-1);
    }
    public void setup()   throws IOException{
        asct.getResponse().getOutputStream().setWriteListener(this);
    }
    private void sendEvent(String data) throws IOException {
        sendEvent(data, "");
    }
    //一个 Event 就是一个字符串
    private void sendEvent(String data, String comment) throws IOException{
        OutboundEvent.Builder builder = new OutboundEvent.Builder();
        OutboundEvent outboundEvent = builder
                .name(comment)
                .data(String.class, data)
```

```java
                .comment(comment)    //目前JavaScript还不能理解comment字段
                .build();
        String str = outboundEvent.toString();
        //采用utf-8对事件信息编码
        InputStream is = new ByteArrayInputStream(str.getBytes("utf-8"));
        //若输出缓冲区满了，则自动冲刷
        is.transferTo(asct.getResponse().getOutputStream());    //从输出流输出
        //.flushBuffer()的作用是把内容冲刷至浏览器，让浏览器收到内容
        //如果不调用flushBuffer()，则浏览器中会出现长时间的空白，然后瞬间出现多条数据
        asct.getResponse().flushBuffer();
    }
    @Override
    public void onWritePossible() throws IOException {
        while (asct.getResponse().getOutputStream().isReady()) {
            try {
                String data = eventSource.getEvent();
                sendEvent(data);
            }catch (InterruptedException ex){
                ex.printStackTrace();
            }
        }
    }
    @Override
    public void onError(Throwable throwable) {
        throwable.printStackTrace();
        asct.complete();
    }
}
```

代码14-4：SseServlet类

```java
public abstract class SseServlet<T extends EventSource> extends HttpServlet {
    //不同的SseServlet应该有不同的SSE_EVENTSOURCE_ATTR
    //以避免同一浏览器访问不同的sse应用产生混淆
    //因此，需要在子类构造器中为此字符串添加当前类名
    private static String SSE_EVENTSOURCE_ATTR = "cxiao.sh.cn.SSEservlet.eventsource";
    private String targetTemplateName;
    protected SseServlet(String templateName){
        this.targetTemplateName = templateName;      //如 "ball/ball"
    }
    protected String getSseEventsourceAttrName(){
        return SSE_EVENTSOURCE_ATTR + "." + this.getClass().getSimpleName();
    }
```

```java
    @Override
    protected void doGet(HttpServletRequest req, HttpServletResponse resp)
                throws ServletException, IOException {
        String path = req.getPathInfo();
        if (path==null){
            setupForm(req, resp);
            return;
        }
        if (path.endsWith("/info")){
            sseInfoFunc(req, resp);
            return;
        }
        T eventSource = (T) req.getSession()
                        .getAttribute(this.getSseEventsourceAttrName());
        if (path.endsWith("/pause")) {
            eventSource.pause();
            return;
        }
        if (path.endsWith("/resume")) {
            eventSource.resume();
        }
    }
    protected abstract T createEventSource();
    private void setupForm(HttpServletRequest req, HttpServletResponse resp)
                    throws ServletException, IOException {
        T eventSource = (T) req.getSession()
                        .getAttribute(this.getSseEventsourceAttrName());
        if (eventSource == null){
            eventSource = createEventSource();
            System.out.println("EventSource ID:" + eventSource.hashCode());
            req.getSession().setAttribute(this.getSseEventsourceAttrName(),
                            eventSource);
        }
        ThymeleafUtil.outputHTML(req, resp, null, targetTemplateName);
    }
    private void sseInfoFunc(HttpServletRequest req, HttpServletResponse resp)
                            throws IOException{
        T eventSource = (T) req.getSession()
                        .getAttribute(this.getSseEventsourceAttrName());
        if (eventSource == null){
            eventSource = createEventSource();
```

```
                req.getSession().setAttribute(this.getSseEventsourceAttrName(),
                                eventSource);
            }
            SSEWriteListener sseWriteListener =
                                new SSEWriteListener(eventSource, req, resp);
            sseWriteListener.setup();
        }
    }
```

第二步：新增如下 ball 应用相关 Servlet 及辅助类。

- Ball。
- TriangleHeap（继承 EventSource）。
- BallServlet（继承 SseServlet）。

代码 14-5：Ball 类

```
@Data
@AllArgsConstructor
public class Ball {
    public static final List<Character> SHAPES =
                        Arrays.asList('■', '▲', '●', '◆', '★');
    public static final int LayersCount = TriangleHeap.LayersCount;
    private int rowIndex = 0;
    private int colIndex = 0;
    private int shapeIndex;
    public Ball(){
        shapeIndex = new Random().nextInt(SHAPES.size());
    }
    public void drop(){
        rowIndex += 1;
        if (rowIndex == LayersCount){
            return;
        }
        int choice = new Random().nextInt(2);
        colIndex += choice;
    }
}
```

代码 14-6：TriangleHeap 类

```
public class TriangleHeap extends EventSource {
    public static final int LayersCount = 12;
    public static final Character DEFAULT_SHAPE = '□';
    private static final long INTERVAL = 200L;
    private int[ ] statistics = new int[LayersCount];
```

```java
private List<Ball> ballList = new ArrayList<>();
//应该把字符串存入消息队列，而不是把TriangleHeap对象存入消息队列
//如果存入后者，则timer在修改TriangleHeap对象的同时sse正在根据此对象构造字符串
//这样会造成访问冲突
private BlockingQueue<String> data = new SynchronousQueue();
record ColumnBar(int colIndex, int count){ }
private ColumnBar columnBar = new ColumnBar(0, 0);
public TriangleHeap(){
    initialize();
}
@Override
protected void initialize() {
    for (int i=0;i<statistics.length;i++){
        statistics[i]=0;
    }
    ballList.add(new Ball());
    timer.scheduleAtFixedRate(this, 0L, INTERVAL);
}
@Override
protected String produceOneEvent() {
    Ball ball;
    if (ballList.size()==1){
        ball = ballList.get(0);
        //初始状态，则直接发送
        if (ball.getColIndex()==0 && ball.getRowIndex()==0){
            try {
                //把对象的字符串形式作为SSE event发送出去
                this.getDataQueue().put(this.toString());
            }catch (InterruptedException ex){
                ex.printStackTrace();
            }
        }
    }
    //从头到尾过一遍，每个球向下移动一层
    for (int i=0;i<ballList.size();i++){
        ball = ballList.get(i);
        ball.drop();
    }
    //当末球行号为LayersCount时，则删除此球
    if (ballList.get(ballList.size()-1).getRowIndex()==LayersCount){
        ball = ballList.remove(ballList.size()-1);
```

```java
            if (statistics[ball.getColIndex()]<100){
                statistics[ball.getColIndex()]+=1;
            }
            this.columnBar =
                new ColumnBar(ball.getColIndex(), statistics[ball.getColIndex()]);
        }
        if (ballList.get(0).getRowIndex()==3){
            ball = new Ball();
            ballList.add(0, ball);
        }
        return this.toString();
    }
    @Override
    public String toString() {
        Character[][] triangle = new Character[LayersCount][LayersCount];
        int i,j;
        for (i=0;i<triangle.length;i++){
            for (j=0;j<=i;j++){
                triangle[i][j] = DEFAULT_SHAPE;
            }
        }
        for (Ball ball:ballList){
            triangle[ball.getRowIndex()][ball.getColIndex()] =
                         Ball.SHAPES.get(ball.getShapeIndex());
        }
        String s = "";
        for (i=0;i<triangle.length;i++){
            for (j=0;j<=i;j++){
                s += triangle[i][j];
            }
            if (i<triangle.length-1){
                s += '*';
            }
        }
        s += ":";
        s += columnBar.colIndex() + "/" + columnBar.count();
        return s;
    }
}
```

代码 14-7：BallServlet 类

```
public class BallServlet extends SseServlet<TriangleHeap> {
    public BallServlet(){
        super("ball/ball");
    }
    @Override
    protected TriangleHeap createEventSource() {
        return new TriangleHeap();
    }
}
```

第三步：在 ServletInitializer 中添加 BallServlet 的注册操作。

代码 14-8：注册 BallServlet

```
//注册 BallServlet
servlet = new BallServlet();
registration = servletContext.addServlet("sse.ball.servlet", servlet);
registration.setLoadOnStartup(8);
registration.addMapping("/ball/*");
registration.setAsyncSupported(true);
```

第四步：在 resources\templates\ 文件夹内添加 ball 应用相关的网页文件。
第五步：在 resources\static\ 文件夹内添加 ball 应用相关的 css 文件及图片文件。
第六步：运行程序，之后在浏览器地址栏中输入 http://localhost:8080/ball，查看显示结果。

14.4 用Spring实现

设计思路

首先增加 Spring 对异步处理的支持，之后在 Controller 提供 SSE 服务的方法内开启一个线程，该线程轮询事件源；当从事件源获得一个事件消息时，则调用 SseEmitter 对象的 send(.)方法把事件发送至客户端。事件源的实现与 Servlet 方案相同。实现步骤如下：

第一步：新增自定义的 SSE 基础类 EventSource，与代码 14-1 相同。
第二步：在 MvcConfig 内重载 configureAsyncSupport(.)方法，增加对异步的支持。

代码 14-9：增加 Spring 对异步的支持

```
@Override
public void configureAsyncSupport(AsyncSupportConfigurer configurer) {
    configurer.setDefaultTimeout(-1);
    configurer.setTaskExecutor(mvcTaskExecutor());
}
@Bean
public ThreadPoolTaskExecutor mvcTaskExecutor() {
    ThreadPoolTaskExecutor taskExecutor = new ThreadPoolTaskExecutor();
```

```
        taskExecutor.setThreadGroupName("mvc-executor");
        return taskExecutor;
    }
```

第三步：新增如下 ball 应用相关类。

❑ Ball 类，与代码 14-5 相同。

❑ TriangleHeap 类，与代码 14-6 相同。

第四步：新增 BallController 类。

代码 14-10：BallController 类

```
@Controller
@RequestMapping(path = "ball")
public class BallController {
    private String SSE_EVENTSOURCE_ATTR =
                        "cxiao.sh.cn.SSE.eventsource.TriangleHeap";
    @Autowired
    private ThreadPoolTaskExecutor taskExecutor;
    @GetMapping
    public String setupForm(HttpSession session){
        getEventSource(session);
        return "ball/ball";
    }
    private TriangleHeap getEventSource(HttpSession session){
        TriangleHeap eventSource =
            (TriangleHeap) session.getAttribute(SSE_EVENTSOURCE_ATTR);
        if (eventSource == null){
            eventSource = new TriangleHeap();
            session.setAttribute(SSE_EVENTSOURCE_ATTR, eventSource);
        }
        return eventSource;
    }
    @GetMapping(path="pause")
    public void pause(HttpServletResponse response, HttpSession session)
                            throws IOException {
        getEventSource(session).pause();
        response.getOutputStream().close();
    }
    @GetMapping(path="resume")
    public void resume(HttpServletResponse response, HttpSession session)
                                throws IOException {
        getEventSource(session).resume();
        response.getOutputStream().close();
    }
    @GetMapping(path="info", produces = {MediaType.TEXT_EVENT_STREAM_VALUE})
```

```java
public SseEmitter push(HttpSession session){
    final SseEmitter emitter = new SseEmitter();
    emitter.onCompletion(()-> System.out.println("完成！"));
    emitter.onError((callback)-> System.out.println("超时！"));
    final TriangleHeap eventSource = getEventSource(session);
    taskExecutor.execute(()->{
        while (true){
            try {
                String data = eventSource.getEvent();
                emitter.send(SseEmitter.event().data(data,
                        MediaType.TEXT_EVENT_STREAM));
            }catch (InterruptedException | IOException ex){
                ex.printStackTrace();
            }
        }
    });
    return emitter;
}
```

第五步：在 resources\templates\ 文件夹内添加 ball 应用相关的网页文件。

第六步：在 resources\static\ 文件夹内添加 ball 应用相关的 CSS 文件及图片文件。

第七步：运行程序，之后在浏览器地址栏中输入 http://localhost:8081/ball，查看显示结果。

14.5 用SpringBoot实现

设计思路

与 Spring 实现方案的设计思路相同。

实现步骤与 14.4 节实现步骤相同。

最后运行程序，在浏览器地址栏中输入 http://localhost:8082/ball，查看显示结果。

14.6 小结

SSE 省去了很多建立连接的过程，在实时性要求较高的场景是一种较好的选择。案例的 Servlet 实现方案在事件源设置一个阻塞队列用于平衡事件产生与消耗的速度；事件的发送放在应答输出流的 WriteListener 内以非阻塞方式输出，避免轮询事件源；用户通过浏览器界面控制事件源内的 Semaphore 对象实现事件生成的暂停与恢复。

14.7 习题

仿照本章案例，在浏览器内显示随机漫步的过程及统计结果。浏览器显示效果如图 14-3 所示,图中黑色方块表示已经走过的位置,白色方块表示尚未走过的位置,箭头表示当前所在位置,箭头的方向表示从哪个方向到达当前位置的,即箭头的反向位置为上一步所在位置。以下是题目描述:某人位于一个 $N\times N$ 区域的中央,他每次随机地选择上、下、左、右四个方向之一并走一步,问要多少步他才能走出这个区域?为了得到更加精确的结果,把他从中央开始到他走出该区域称为一次模拟,程序做 K 次模拟,计算这 K 次模拟的平均步数。

图14-3 浏览器显示效果

第 15 章
WebSocket

本章介绍如何使用 Servlet 处理 WebSocket 通信。本章案例是编写一个简易聊天室。
通过学习本章内容，读者将可以：
❏ 用 Servlet 开发 WebSocket 应用

15.1 相关概念

WebSocket 使得客户端和服务器之间的数据交换变得简单，允许服务端主动向客户端推送数据。在 WebSocket API 中，客户端和服务端只需要完成一次握手，两者之间就直接创建持久性的连接，可以进行双向数据传输。

15.1.1 WebSocket概述

WebSocket 协议（RFC 6455）消除了 HTTP 连接的无状态特性，提供了在一条 TCP 信道上客户端和服务端之间全双工的通信。图 15-1 显示了 WebSocket 与 HTTP 在通信上的不同。

Jakarta-websocket-spec 是 WebSocket 服务的 Java API 规范，是使用 Java 开发 WebSocket 服务时的基本约定。

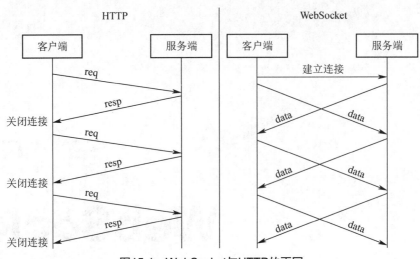

图15-1　WebSocket与HTTP的不同

15.1.2　事件驱动API

Jakarta-websocket-spec 定义了表 15-1 所列的注解，这些注解用于标注 WebSocket 的事件处理函数。当注解所对应的事件发生时，注解所标注的函数会自动执行，事件的相关信息将作为参数传入函数。所标注的函数又称回调函数。

表 15-1　事件处理函数的注解

注　解	对　应　事　件	函数参数约定
@OnOpen	当连接建立时	❑ 一个可选的 Session 参数 ❑ 一个可选的 EndpointConfig 参数 ❑ 0~n 个标注了@PathParam 的（String 类型、Java 简单类型或其装箱类）参数
@OnMessage	当接收到消息时	❑ 一个可选的 Session 参数 ❑ 一个必需的表示所接收信息（文本信息/参数类型 String、二进制信息/参数类型 byte[ ]、pong 信息/参数类型 PongMessage）的参数 ❑ 0~n 个标注了@PathParam 的（String 类型、Java 简单类型或其装箱类）参数 ❑ 如果@OnMessage 标注的方法有返回值，这个返回值将作为消息发送给通信的对方
@OnError	当连接出错时	❑ 一个可选的 Session 参数 ❑ 一个必需的 Throwable 参数 ❑ 0~n 个标注了@PathParam 的(String 类型、Java 简单类型或其装箱类)参数
@OnClose	当关闭连接时	❑ 一个可选的 Session 参数 ❑ 一个可选的 CloseReason 参数 ❑ 0~n 个标注了@PathParam 的（String 类型、Java 简单类型或其装箱类）参数

上述注解应该在标注了@ServerEndpoint 或者@ClientEndpoint 的 Java 类中使用。

15.2 案例描述

多个用户访问聊天室进行聊天，界面效果如图 15-2 所示。

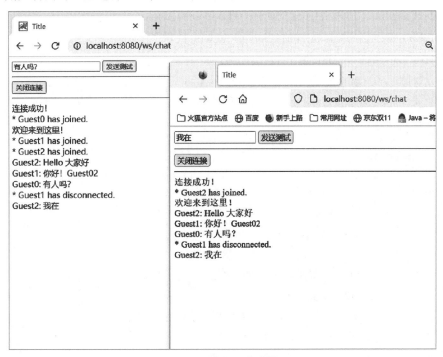

图15-2 案例运行界面

15.3 用Servlet实现

设计思路

创建标注了@ServerEndpoint 的 TalkEndpoint 类，用于接收从 JavaScript 发出的 WebSocket 请求并向 JavaScript 应答，在 TalkEndpoint 类中维护一个连接集，以便聊天室里一个用户的发言内容可以发送至所有用户的窗口中；创建 WebSocketServlet，用于在浏览器窗口显示初始界面。实现步骤如下：

第一步：在模块的 pom.xml 中添加 WebSocket 依赖。

代码 15-1：WebSocket 依赖

```
<dependency>
    <groupId>jakarta.websocket</groupId>
    <artifactId>jakarta.websocket-api</artifactId>
    <version>2.1.0</version>
</dependency>
<dependency>
    <groupId>org.apache.tomcat.embed</groupId>
```

```xml
    <artifactId>tomcat-embed-websocket</artifactId>
    <version>${tomcat.version}</version>
</dependency>
```

第二步：增加 WebSocket 的 Endpoint 类。

- IndexEndpoint（使用 Jakarta 的@ServerEndpoint 注解）
- TalkEndPoint（使用 Jakarta 的@ServerEndpoint 注解）

代码 15-2：IndexEndpoint 类

```java
//此注解的 value 指明 websocket 协议访问的 URI，在 JavaScript 中通过
//ws://host:port/index 访问此 URI 建立持久连接，不能通过浏览器地址栏访问此 URI
@ServerEndpoint(value = "/index")
public class IndexEndpoint {
    @OnMessage
    public String messageReceiver(String message) {
        System.out.println("websocket 收到信息: " + message);
        return "Message Received: " + message;
    }
}
```

IndexEndpoint 类与聊天室无关，作用是对比。

代码 15-3：TalkEndpoint 类

```java
//TalkEndpoint 类似于 Servlet，不同的是 Servlet 是单例
//而 TalkEndPoint 是每个连接对应其一个实例
@ServerEndpoint(value = "/talk/{page}")
public class TalkEndpoint {
    private static final String GUEST_PREFIX = "Guest";
    //记录当前有多少个用户加入到了聊天室，它是 static 全局变量
    //为了线程安全，使用原子变量 AtomicInteger
    private static final AtomicInteger connectionIds = new AtomicInteger(0);
    //每个用户用一个 TalkEndPoint 实例来维护
    //注意这是一个全局的 static 变量，所以用到了线程安全的 CopyOnWriteArraySet
    private static final Set<TalkEndpoint> connections = new CopyOnWriteArraySet<>();
    private String nickname;
    private Session session;          //代表底层的通信连接
    public TalkEndpoint(){
        nickname = GUEST_PREFIX + connectionIds.getAndIncrement();
        System.out.println("创建一个 EndPoint 对象, nickname=" + nickname);
    }
    //新连接建立时，Tomcat 会创建一个 Session，并回调该函数
    @OnOpen
    public void open(@PathParam("page") String page, Session session) {
        System.out.println("服务器建立连接: " + page);
        this.session = session;
```

```java
        connections.add(this);
        String message = String.format("* %s %s", nickname, "has joined.");
        broadcast(message);
    }
    //浏览器关闭连接时，Tomcat会回调这个函数
    @OnClose
    public void close(@PathParam("page") String page, Session session){
        System.out.println("服务器关闭连接: " + page);
        connections.remove(this);
        String message = String.format("* %s %s", nickname, "has disconnected.");
        broadcast(message);
    }
    //浏览器发送消息到服务器时，Tomcat会回调这个函数
    @OnMessage
    public void receiveMessage(@PathParam("page") String page,
                               Session session,
                               String message)
                            throws IOException {
        System.out.println(String.format(page + "接受到用户" +
                                session.getId() + "的数据:" + message));
        String filteredMessage = String.format("%s: %s", nickname, message);
        broadcast(filteredMessage);
        //如果该函数返回String，则当前客户端会收到此返回字符串
        //return message;
    }
    //WebSocket连接出错时，Tomcat会回调这个函数
    @OnError
    public void error(Throwable throwable){
        try {
            throw throwable;
        } catch (Throwable e) {
            System.out.println("未知错误");
        }
    }
    //向聊天室中的每个用户广播消息
    private static void broadcast(String msg) {
        for (TalkEndpoint client : connections) {
            try {
                synchronized (client) {
                    client.session.getBasicRemote().sendText(msg);
                }
            } catch (IOException e) {
```

```
                e.printStackTrace();
            }
        }
    }
}
```

第三步：增加 WebSocket 的用户界面类 WebSocketServlet。

代码 15-4：WebSocketServlet 类

```
public class WebSocketServlet extends HttpServlet {
    @Override
    protected void doGet(HttpServletRequest req,
                    HttpServletResponse resp)
                    throws ServletException, IOException {
        String path = req.getPathInfo();
        if (path==null) {
            ThymeleafUtil.outputHTML(req, resp, null, "ws/indexPage");
        }else{
            ThymeleafUtil.outputHTML(req, resp, null, "ws/chatPage");
        }
    }
}
```

第四步：在 ServletInitializer 类中添加对 WebSocketServlet 的注册操作。

代码 15-5：注册 WebSocketServlet

```
//注册 WebSocketServlet
servlet = new WebSocketServlet();
registration = servletContext.addServlet("websocket.serlvet", servlet);
registration.setLoadOnStartup(9);
registration.addMapping("/ws/*");
registration.setAsyncSupported(true);
```

第五步：在 resources\templates\ws\文件夹内添加与 WebSocketServlet 有关的网页文件

❑ indexPage.html
❑ chatPage.html

第六步：运行程序，之后在浏览器地址栏中输入 http://localhost:8080/ws 以及 http://localhost:8080/ws/chat，查看显示结果。

15.4 用Spring实现

设计思路

创建 TextWebSocketHandler 的派生类并重载其方法。这些方法是回调函数，在 WebSocket 相应的事件发生时自动执行。这些方法都有一个 WebSocketSession 类型的参数，此参数代表 WebSocket 连接，通过此连接可以主动发送信息。信息的接收则是被动触发的，无须主动执行接收信息的操作。WebSocket 的 URI 在 WebSocketConfigurer 接口的实现类中配置。实现步骤如下：

第一步：在模块的 pom.xml 中添加 WebSocket 依赖。

代码 15-6：WebSocket 依赖

```xml
<dependency>
    <groupId>org.springframework</groupId>
    <artifactId>spring-websocket</artifactId>
    <version>${spring.version}</version>
</dependency>
<dependency>
    <groupId>org.apache.tomcat.embed</groupId>
    <artifactId>tomcat-embed-websocket</artifactId>
    <version>${tomcat.version}</version>
</dependency>
```

第二步：增加 WebSocket 的 Endpoint 类。

❏ IndexEndpoint（继承 Spring 的 TextWebSocketHandler 类）
❏ TalkEndpoint（继承 Spring 的 TextWebSocketHandler 类）

代码 15-7：IndexEndpoint 类

```java
@Component
public class IndexEndpoint extends TextWebSocketHandler {
    @Override
    protected void handleTextMessage(WebSocketSession session,
                                     TextMessage message)
                                     throws Exception {
        String msg = message.getPayload();
        session.sendMessage(new TextMessage("websocket 收到信息: " + msg));
    }
}
```

IndexEndpoint 类与聊天室无关，作用是对比。

代码 15-8：TalkEndpoint 类

```java
@Component
public class TalkEndpoint extends TextWebSocketHandler {    //单例
    private static final String GUEST_PREFIX = "Guest";
    //记录当前有多少个用户加入到了聊天室，它是 static 全局变量
    //为了线程安全，使用原子变量 AtomicInteger
    private static final AtomicInteger connectionIds = new AtomicInteger(0);
    //注意这是一个全局的 static 变量，所以用到了线程安全的 CopyOnWriteArraySet
    private static final Set<WebSocketSession> connections =
                                    new CopyOnWriteArraySet<>();
    @Override
    public void afterConnectionEstablished(WebSocketSession session)
```

```java
    throws Exception {
        System.out.println("服务器建立连接...");
        connections.add(session);
        String nickname = GUEST_PREFIX + connectionIds.getAndIncrement();
        Map<String,Object> map = session.getAttributes();
        map.put("nickname", nickname);
        String message = String.format("* %s %s", nickname, "has joined.");
        broadcast(message);
    }
    @Override
    protected void handleTextMessage(WebSocketSession session,
                                    TextMessage message)
                                    throws Exception {
        var msg = message.getPayload();
        System.out.println(String.format("接收到连接" + session.getId() +
                                        "的数据:" + msg));
        String nickname = (String) session.getAttributes().get("nickname");
        String filteredMessage = String.format("%s: %s", nickname, msg);
        broadcast(filteredMessage);
    }
    @Override
    public void handleTransportError(WebSocketSession session,
                                    Throwable exception)
                                    throws Exception {
        try {
            throw exception;
        } catch (Throwable e) {
            System.out.println("未知错误");
        }
    }
    @Override
    public void afterConnectionClosed(WebSocketSession session,
                                    CloseStatus status)
                                    throws Exception {
        System.out.println(session.getId() + "服务端关闭连接。");
        connections.remove(session);
        String nickname = (String) session.getAttributes().get("nickname");
        String message = String.format("* %s %s", nickname, "has disconnected.");
        broadcast(message);
    }
    private static void broadcast(String msg) {
        for (WebSocketSession client : connections) {
            try {
```

```
            synchronized (client) {
                //这里可以多次调用 sendMessage(.)方法
                client.sendMessage(new TextMessage(msg));
            }
        } catch (IOException e) {
            e.printStackTrace();
        }
    }
  }
}
```

第三步：新增配置类 WebSocketConfig（实现 Spring 的 WebSocketConfigurer 接口），对 WebSocket URI 进行配置。

代码 15-9：WebSocketConfig 类

```
@EnableWebSocket
@Configuration
public class WebSocketConfig implements WebSocketConfigurer {
    @Resource
    TextWebSocketHandler indexEndpoint;
    @Resource
    TextWebSocketHandler talkEndpoint;
    @Override
    public void registerWebSocketHandlers(WebSocketHandlerRegistry registry) {
        //等同于 Servlet 实现方案中的 @ServerEndpoint(value = "/index")
        registry.addHandler(indexEndpoint, "/index");
        //等同于 Servlet 实现方案中的 @ServerEndpoint(value = "/talk")
        registry.addHandler(talkEndpoint, "/talk");
    }
}
```

第四步：增加 WebSocket 的用户界面类 WebSocketController。

代码 15-10：WebSocketController 类

```
@Controller
@RequestMapping(path="ws")
public class WebSocketController {
    @GetMapping(path="/{action}")
    public String setup(@PathVariable("action") String action){
        if (action.equals("index")){
            return "ws/indexPage";
        }
        if (action.equals("chat")){
            return "ws/chatPage";
        }
```

```
        return "";
    }
    @GetMapping             //复用上述setup(.)方法
    public String setup(){
        return setup("index");
    }
}
```

第五步：在 resources\templates\ws\ 文件夹内添加与 WebSocketController 有关的网页文件。

- indexPage.html
- chatPage.html

第六步：运行程序，之后在浏览器地址栏中输入 http://localhost:8081/ws 以及 http://localhost:8081/ws/chat，查看显示结果。

15.5　用SpringBoot实现

设计思路

与 Spring 实现方案的设计思路相同。实现步骤如下：

第一步：在模块的 pom.xml 中添加 WebSocket 依赖。

代码 15-11：WebSocket 依赖

```
<dependency>
    <groupId>org.springframework.boot</groupId>
    <artifactId>spring-boot-starter-websocket</artifactId>
    <version>${springboot.version}</version>
</dependency>
```

第二步：增加 WebSocket 的 Endpoint 类，与 15.4 节第二步相同。

第三步：新增配置类 WebSocketConfig（实现 Spring 的 WebSocketConfigurer 接口），对 WebSocket URI 进行配置，与 15.4 节第三步相同。

第四步：增加 WebSocket 的用户界面类 WebSocketController，与 15.4 节第四步相同。

第五步：在 resources\templates\ws\ 文件夹内添加与 WebSocketController 有关的网页文件，与 15.4 节第五步相同。

第六步：运行程序，之后在浏览器地址栏中输入 http://localhost:8082/ws 以及 http://localhost:8082/ws/chat，查看显示结果。

15.6　小结

WebSocket 是一种全新的协议，协议名为 "ws"，不属于 HTTP 无状态协议。客户端用形如 ws://host:port/<uri> 的地址访问 value 值设为 <uri> 的 @ServerEndpoint 注解所标注的类，每个

WebSocket 连接在服务端产生该类的一个实例。在定义 WebSocket 端点对象时，用 Jakarta-websocket-spec 所规定的@OnOpen、@OnMessage、@OnClose、@OnError 注解标注函数，所标注的函数在对应的事件发生时会自动执行。Servlet 本身并不能提供 WebSocket 服务，但是 Servlet 应答的网页内有 JavaScript 可以访问 WebSocket 服务。

15.7 习题

用 WebSocket 实现 14.7 节习题。

第 16 章
异常的统一处理

本章介绍如何对 Servlet 的运行异常进行统一处理。本章案例是用统一的页面显示 Servlet 运行时出现的异常。

通过学习本章内容，读者将可以：
- 对 Servlet 的异常进行统一处理

16.1 相关概念

Servlet 或过滤器在处理请求时可能抛出运行时异常、ServletException 或其子类对象、IOException 或其子类对象。这些异常可以交由某个 Servlet 统一处理。

16.1.1 sendError(.)方法

HttpServletResponse 的 sendError(.)方法把参数作为出错信息，为其设置相应的消息头和消息体，并发送至客户端。此方法执行后会提交并终结应答。执行此方法后不应该再向客户端输出数据，因为这些数据将被忽略。

调用 sendError(.)方法时，如果在此方法调用之前写入 ServletResponse 缓冲区的数据尚未被发送至客户端，此方法将删除缓冲区内已有的数据，并写入方法所设置的数据；如果在调用此方法之前应答已经被提交，此方法会抛出 IllegalStateException 异常。

16.1.2 出错处理Servlet

Servlet 容器可以把 Servlet 产生的错误交由指定的 Servlet（称为出错处理 Servlet）来处理。相关出错信息通过 ServletRequest 的相关命名属性传递给出错处理 Servlet，见表 16-1。

表 16-1 传递出错信息的命名属性

命 名 属 性	数 据 类 型
jakarta.servlet.error.status_code	java.lang.Integer
jakarta.servlet.error.exception_type	java.lang.Class
jakarta.servlet.error.message	java.lang.String
jakarta.servlet.error.exception	java.lang.Throwable
jakarta.servlet.error.request_uri	java.lang.String
jakarta.servlet.error.servlet_name	java.lang.String

出错处理 Servlet 读取这些命名属性，根据状态码、异常类型、出错信息、异常对象、出错时的请求 URI 以及出错 Servlet 的逻辑名称来产生定制的应答内容并发送至客户端。

调用 sendError(.)方法亦会引发出错处理 Servlet 的执行。

16.2 案例描述

图 16-1 所示为 Servlet 抛出异常时定制的浏览器显示效果，图 16-2 所示为执行 sendError(.)方法时定制的浏览器显示效果。

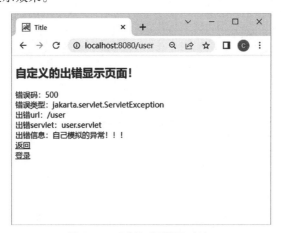

图16-1 案例运行界面（1）

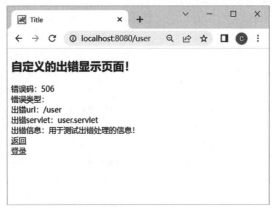

图16-2 案例运行界面（2）

16.3 用Servlet实现

设计思路

创建 ErrorServlet，用于显示出错信息。向 ServletContext 注册 ErrorServlet 并配置 URI，这个 ErrorServlet 并不是通过浏览器地址栏访问，而是在设置 Servlet 容器的 ErrorPage 时指定该 ErrorServlet，这样在调用 sendError(.)方法时或者在 Servlet 容器中出现未捕获异常时会跳转至此 ErrorServlet 处理。实现步骤如下：

第一步：新增 ErrorServlet 类，用于处理系统运行抛出的异常。

代码 16-1：ErrorServlet 类

```java
public class ErrorServlet extends HttpServlet {
    @Override
    protected void doGet(HttpServletRequest req,
                         HttpServletResponse resp)
                    throws ServletException, IOException {
        Map<String, Object> errors = new HashMap<String, Object>();
        Throwable throwable = (Throwable)
                        req.getAttribute("jakarta.servlet.error.exception");
        if (throwable!=null){       //由 Servlet 抛出异常引起的出错处理
            errors.put("jakarta.servlet.error.exception",
                    throwable.getClass().getName());
            errors.put("jakarta.servlet.error.message",
                    throwable.getMessage());
        }else{                      //由 sendError(.)引起的出错处理
            errors.put("jakarta.servlet.error.exception",
                        req.getAttribute("jakarta.servlet.error.exception_type"));
            errors.put("jakarta.servlet.error.message",
                        req.getAttribute("jakarta.servlet.error.message"));
        }
        errors.put("jakarta.servlet.error.status_code",
                        req.getAttribute("jakarta.servlet.error.status_code"));
        errors.put("jakarta.servlet.error.request_uri",
                        req.getAttribute("jakarta.servlet.error.request_uri"));
        errors.put("jakarta.servlet.error.servlet_name",
                        req.getAttribute("jakarta.servlet.error.servlet_name"));
        //出错页面应该直接显示，不宜分派给/display Servlet 去显示
        ITemplateEngine templateEngine = (ITemplateEngine) req.getServletContext()
            .getAttribute("cn.shanghai.cxiao.thymeleaf.TemplateEngineInstance");
        IWebExchange webExchange = JakartaServletWebApplication
            .buildApplication(req.getServletContext()).buildExchange(req, resp);
        WebContext context = new WebContext(webExchange, req.getLocale());
        context.setVariable("errors", errors);
```

```
            templateEngine.process("exception/error", context, resp.getWriter());
    }
}
```

第二步：添加 ErrorServlet 对应的网页 error.html 及静态文件。

代码 16-2：error.html

```
<!DOCTYPE html>
<html xmlns:th="http://www.thymeleaf.org">
<head>
    <meta charset="UTF-8">
    <title>Title</title>
</head>
<body>
<h2>自定义的出错显示页面！</h2>
错误码：<span th:text="${errors.get('jakarta.servlet.error.status_code')}">
</span><br>
错误类型：<span th:text="${errors.get('jakarta.servlet.error.exception')}">
</span><br>
出错 url：<span th:text="${errors.get('jakarta.servlet.error.request_uri')}">
</span><br>
出错 servlet：<span th:text="${errors.get('jakarta.servlet.error.servlet_name')}">
</span><br>
出错信息：<span th:text="${errors.get('jakarta.servlet.error.message')}">
</span><br>
<a th:href="@{${errors.get('jakarta.servlet.error.request_uri')}}">返回
</a> <br>
<a href="/">登录</a>
</body>
</html>
```

第三步：在 ServletInitializer 中注册 ErrorServlet，为其指定 URI。

代码 16-3：注册 ErrorServlet

```
//注册 ErrorServlet
servlet = new ErrorServlet();
registration = servletContext.addServlet("error.servlet", servlet);
registration.setLoadOnStartup(10);
registration.addMapping("/error");
registration.setAsyncSupported(true);
```

第四步：在 Tomcat 启动之前设置出错时跳转的 URI，此 URI 已在第三步映射至 ErrorServlet。

代码 16-4：设置 Servlet 容器的出错处理功能

```
//设置出错页面，当出现错误时，将转向特定的 Servlet 处理。
ErrorPage errorPage = new ErrorPage();
//已经设置了 uri 为/error 的 ErrorServlet
```

```
errorPage.setLocation("/error");
//可设置出错页面只针对特定错误号或特定异常类型
//errorPage.setErrorCode(500);
//errorPage.setExceptionType("jakarta.servlet.ServletException");
context.addErrorPage(errorPage);
```

第五步：在 UserServlet 的 doGet(.)方法中增加出错的测试代码。

代码 16-5：测试代码

```
if(true) {
    resp.sendError(506, "用于测试出错处理的信息！");
    return;
    //或者使用下面的语句
    //throw new ServletException("自己模拟的异常!!! ");
}
```

第六步：运行系统，在浏览器地址栏中输入 http://localhost:8080/user，查看显示结果。

16.4 用Spring实现

设计思路

创建实现 HandlerExceptionResolver 接口的 Bean，通过该 Bean 设置出错时显示的页面，再向 Spring 注册这个 Bean。实现步骤如下：

第一步：在 MvcConfig 中创建实现 HandlerExceptionResolver 接口的 Bean，该 Bean 可以对指定类型的未捕获异常进行处理。

代码 16-6：HandlerExceptionResolver Bean

```
@Bean
public HandlerExceptionResolver handlerExceptionResolver() {
    SimpleMappingExceptionResolver exceptionResolver =
                            new SimpleMappingExceptionResolver();
    //设置默认的出错时显示的 HTML 网页
    exceptionResolver.setDefaultErrorView("exception/error");
    //设置针对某些特定异常的出错显示页面
    Properties exceptionMapping = new Properties();
    exceptionMapping.setProperty(DataReadException.class.getName(),
                        "exception/dberror");
    exceptionResolver.setExceptionMappings(exceptionMapping);
    return exceptionResolver;
}
```

第二步：在 MvcConfig 中重载 configureHandlerExceptionResolvers(.)方法，向 Spring 注册在第一步创建的 HandlerExceptionResolver Bean。

代码 16-7：configureHandlerExceptionResolvers(.)方法

```
@Override
```

```
public void configureHandlerExceptionResolvers(
        List<HandlerExceptionResolver> resolvers) {
    resolvers.add(handlerExceptionResolver());
}
```

第三步：新增在 HandlerExceptionResolver Bean 中指定了 View name 的网页，如 resources\templates\文件夹内的 exception\error.html。

代码 16-8：error.html

```
<!DOCTYPE html>
<html xmlns:th="http://www.thymeleaf.org">
<head>
    <meta charset="UTF-8">
    <title>Title</title>
</head>
<body>
<h2>自定义的出错显示页面！</h2>
错误信息：<span th:text="${exception}"></span><br>
<a href="/">登录</a>
</body>
</html>
```

在 HTML 页面中用${exception}引用出错信息。

第四步：在 UserController 中增加出错的测试代码。

代码 16-9：测试代码

```
if(true) {
    throw new ServletException("自己模拟的异常!!! ");
}
```

第五步：运行系统，在浏览器地址栏中输入 http://localhost:8081/user，查看显示结果。

16.5 用SpringBoot实现

设计思路

SpringBoot 的异常处理最为简单，只需要在 application.properties 内增加 server.error.XXXX 配置，并提供 error.html 页面。实现步骤如下：

第一步：在 application.properties 内增加配置。

代码 16-10：配置 server.error.XXX

```
# 是否使用whitelabel出错页面，默认值为true
server.error.whitelabel.enabled=false
# 异常名称是否包含在model内，默认值为false
server.error.include-exception=true
# 异常的stacktrace是否包含在model内，默认值为never
server.error.include-stacktrace=always
```

第二步：在 resources\templates\文件夹内添加 error.html 网页，位置及文件名不能改变。

代码 16-11：error.html

```
<!DOCTYPE html>
<html xmlns:th="http://www.thymeleaf.org">
<head>
    <meta charset="UTF-8">
    <title>Title</title>
</head>
<body>
    <h2>自定义的出错显示页面！</h2>
    <div>
        <div>
            <span><strong>Status</strong></span>
            <span th:text="${status}"></span>
        </div>
        <div>
            <span><strong>Error</strong></span>
            <span th:text="${error}"></span>
        </div>
        <div>
            <span><strong>Message</strong></span>
            <span th:text="${message}"></span>
        </div>
        <div th:if="${exception != null}">
            <span><strong>Exception</strong></span>
            <span th:text="${exception}"></span>
        </div>
        <div th:if="${trace != null}">
            <h3>Stacktrace</h3>
            <span th:text="${trace}"></span>
        </div>
    </div>
    <a href="/">登录</a>
</body>
</html>
```

根据 SpringBoot 的约定，在 HTML 页面中可以直接引用${status}等出错信息。

第三步：在 UserController 中增加出错的测试代码。

代码 16-12：测试代码

```
if(true) {
    throw new ServletException("自己模拟的异常!!! ");
}
```

第四步：运行系统，在浏览器地址栏中输入 http://localhost:8082/user，查看显示结果。

16.6 小结

如果 Servlet 或过滤器在处理请求时发生异常而这些异常未被捕获，应该给浏览器一个反馈页面告知出现了错误。而页面只有通过 Servlet 才能发送出去，这个 Servlet 就是自定义的 ErrorServlet。ErrorServlet 只有通过 ServletContext 的 addErrorPage(.)方法设置才具有对异常统一处理的能力。异常信息通过规定的 ServletRequest 命名属性传递给 ErrorServlet，ErrorServlet 再把这些异常信息显示在反馈页面内。如果不是抛出异常而是在代码中显式调用 HttpServletResponse 的 sendError(.)方法，那么也会引发 ErrorServlet 的处理。

16.7 习题

创建不同的出错处理 Servlet，分别处理不同类型的异常，并用代码测试。

附录 A
安装及设置

1. jdk

安装 JDK17，设置环境变量 JAVA_HOME 为 jdk 安装路径；之后在环境变量 Path 中添加条目 %JAVA_HOME%\bin\。

2. maven

安装 maven 之后，假设安装路径为 D:\maven-3.6.1，在安装路径下 conf\settings.xml 中进行如下配置：

（1）设置 localRepository。

代码 附 A-1：设置 maven 的 localRepository

```
<localRepository>D:\maven-3.6.1\LocalRepository</localRepository>
```

（2）在 `<mirrors>...</mirrors>` 中添加镜像仓库。

代码 附 A-2：添加 maven 的镜像仓库

```
<mirror>
  <id>alimaven</id>
  <mirrorOf>central</mirrorOf>
  <name>aliyun maven</name>
  <url>http://maven.aliyun.com/nexus/content/repositories/central/</url>
</mirror>
```

（3）在 `<profiles>...</profiles>` 中添加 profile。

代码 附 A-3：添加 maven 的 profile

```
<profile>
    <id>jdk-17</id>
    <activation>
        <activeByDefault>true</activeByDefault>
```

```xml
        <jdk>17</jdk>
    </activation>
    <properties>
        <project.build.sourceEncoding>UTF-8</project.build.sourceEncoding>
        <maven.compiler.encoding>UTF-8</maven.compiler.encoding>
        <java.version>17</java.version>
        <maven.compiler.source>17</maven.compiler.source>
        <maven.compiler.target>17</maven.compiler.target>
        <maven.compiler.compilerVersion>17</maven.compiler.compilerVersion>
    </properties>
</profile>
```

3. IDEA 设置

IDEA 通常要作两处设置。一处是设置 maven，设置界面如图附 A-1 所示。

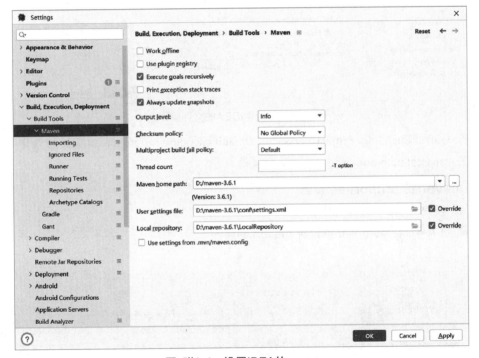

图 附A-1　设置IDEA的maven

说明：勾选 Always update snapshots 复选框，取消勾选 Use settings from .mvn/maven.config 复选框，再根据 maven 的安装路径设置 Maven home path、User settings file、Local repository。

另一处是设置文件编码，可统一设置成 UTF-8 编码，设置界面如图附 A-2 所示。

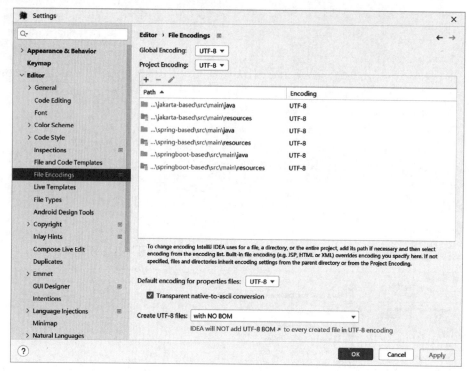

图 附A-2　设置IDEA的文件编码

说明：Global Encoding、Project Encoding、Default encoding for properties files 均选择 UTF-8，再勾选 Transparent native-to-ascii conversion 复选框。

4. maven 的 properties 设置

在每个模块的 pom.xml 文件中添加 maven 的属性设置。

代码 附 A-4：在 pom.xml 中设置 maven 的属性

```xml
<properties>
    <project.build.sourceEncoding>UTF-8</project.build.sourceEncoding>
    <java.version>17</java.version>
    <maven.compiler.encoding>UTF-8</maven.compiler.encoding>
    <maven.compiler.source>17</maven.compiler.source>
    <maven.compiler.target>17</maven.compiler.target>

    <tomcat.version>10.1.7</tomcat.version>
    <jackson.version>2.14.2</jackson.version>
</properties>
```

5. 项目打包的 maven 配置

maven 有两种打包方式，胖包与瘦包。

方式一：胖包，使用 maven-shade-plugin，将项目的依赖以及项目的源码打包成一个可执行 Jar 包。推荐使用这种打包方式。

代码 附 A-5:maven 胖包的 maven-shade-plugin 配置

```xml
<plugins>
    <plugin>
        <groupId>org.apache.maven.plugins</groupId>
        <artifactId>maven-shade-plugin</artifactId>
        <version>3.3.0</version>
        <executions>
            <execution>
                <phase>package</phase>
                <goals>
                    <goal>shade</goal>
                </goals>
                <configuration>
                    <transformers>
                        <!-- 指定启动类 -->
                        <transformer implementation=
"org.apache.maven.plugins.shade.resource.ManifestResourceTransformer">
                            <mainClass>cxiao.sh.cn.MyApp</mainClass>
                        </transformer>
                    </transformers>
                </configuration>
            </execution>
        </executions>
    </plugin>
</plugins>
```

另一种打成胖包的方法是使用 maven-assembly-plugin,用此 plugin 也可以得到一个包含项目的依赖及项目的源码并且可执行的 Jar 包,配置如下:

代码 附 A-6:maven 胖包的 maven-assembly-plugin 配置

```xml
<plugins>
    <plugin>
        <groupId>org.apache.maven.plugins</groupId>
        <artifactId>maven-assembly-plugin</artifactId>
        <version>3.3.0</version>
        <configuration>
            <archive>
                <!-- 指定启动类 -->
                <manifest>
                    <mainClass>cxiao.sh.cn.MyApp</mainClass>
                </manifest>
            </archive>
            <!-- 描述后缀 -->
            <descriptorRefs>
```

```xml
                <descriptorRef>jar-with-dependencies</descriptorRef>
            </descriptorRefs>
        </configuration>
        <!-- 相当于在执行 package 打包时，在后面加上 assembly:single   -->
        <executions>
            <execution>
                <id>make-assembly</id>
                <phase>package</phase>
                <goals>
                    <goal>single</goal>
                </goals>
            </execution>
        </executions>
    </plugin>
</plugins>
```

方式二：瘦包，使用 maven-jar-plugin 和 maven-dependency-plugin（两者缺一不可）。

maven-jar-plugin 把项目源码（不含依赖）打包，在包中指定启动类、指定依赖包相对于项目最终 Jar 包所在的路径、给 MANIFEST.MF 文件添加 Class-Path 属性，以便运行项目 Jar 包时会根据 Class-Path 属性找到其他所依赖 Jar 包的路径。

maven-dependency-plugin 把项目所依赖的各个 Jar 包复制到指定的文件夹内，这个文件夹在配置 maven-jar-plugin 时会被引用。

代码 附 A-7：maven 瘦包的配置

```xml
<plugins>
    <plugin>
        <groupId>org.apache.maven.plugins</groupId>
        <artifactId>maven-dependency-plugin</artifactId>
        <version>3.3.0</version>
        <executions>
            <execution>
                <id>copy-dependencies</id>
                <phase>package</phase>
                <goals>
                    <goal>copy-dependencies</goal>
                </goals>
                <configuration>
                    <!--复制项目依赖包到 lib/目录下 -->
                    <outputDirectory>
                        ${project.build.directory}/lib
                    </outputDirectory>
                    <!--间接依赖也复制-->
                    <excludeTransitive>false</excludeTransitive>
                    <!--带上版本号-->
```

```xml
                    <stripVersion>false</stripVersion>
                </configuration>
            </execution>
        </executions>
    </plugin>
    <plugin>
        <groupId>org.apache.maven.plugins</groupId>
        <artifactId>maven-jar-plugin</artifactId>
        <version>3.3.0</version>
        <configuration>
            <archive>
                <addMavenDescriptor>true</addMavenDescriptor>
                <manifest>
                    <useUniqueVersions>false</useUniqueVersions>
                    <addClasspath>true</addClasspath>
                    <!--这项要与maven-dependency-plugin中的位置一致-->
                    <classpathPrefix>lib/</classpathPrefix>
                    <mainClass>cxiao.sh.cn.MyApp</mainClass>
                </manifest>
            </archive>
            <includes>
                <include>**/*.*</include>
            </includes>
        </configuration>
    </plugin>
</plugins>
```

附录 B
初始项目

源码中项目 chapter00 是本书的起点项目，包括三个模块。

1. jakarta-based 模块

设计思路

用代码创建 Tomcat 服务器对象，配置 Web 服务端口，再启动 Web 服务。实现步骤如下：

第一步：在 pom.xml 中添加嵌入式 tomcat 的依赖。

代码 附 B-1：tomcat 的依赖

```xml
<dependency>
    <groupId>org.apache.tomcat.embed</groupId>
    <artifactId>tomcat-embed-core</artifactId>
    <version>${tomcat.version}</version>
</dependency>
```

第二步：创建 TOMCAT 类，指定服务端口为 8080。

代码 附 B-2：TOMCAT 类

```java
public class TOMCAT {
    private static int PORT = 8080;
    public static void service() throws Exception{
        //创建 Tomcat 对象
        Tomcat tomcat = new Tomcat();
        //设置服务端口
        Connector connector = new Connector();
        connector.setPort(PORT);
        connector.setURIEncoding("UTF-8");
        tomcat.getService().addConnector(connector);
        //启动 Web 服务
        tomcat.start();
```

```
        System.out.println("Tomcat starting...");
        tomcat.getServer().await();
    }
}
```

第三步：添加入口函数

代码 附 B-3：main(.)函数

```
public class MyApp {
    public static void main(String[ ] args) throws Exception {
        TOMCAT.service();
    }
}
```

2. spring-based 模块

设计思路

设计思路与 jakarta-base 模块相同。

模块的实现步骤与 jakarta-base 模块的实现步骤相同。为了在浏览器上有所区分，把 spring-based 模块的服务端口改成 8081。对应的代码为：

```
private static int PORT = 8081;
```

3. springboot-based 模块

设计思路

SpringBoot 的自动配置功能使得只需要设置 Web 服务端口，无须显式启动 Web 服务。实现步骤如下：

第一步：在 pom.xml 中添加嵌入式 tomcat 的依赖。

代码 附 B-4：tomcat 的依赖

```
<dependency>
    <groupId>org.springframework.boot</groupId>
    <artifactId>spring-boot-starter-web</artifactId>
    <version>${springboot.version}</version>
</dependency>
```

第二步：在 resources\application.properties 文件中设置服务端口为 8082。

```
server.port=8082
```

第三步：添加入口函数。

代码 附 B-5：main(.)函数

```
@SpringBootApplication
public class MyApp {
    public static void main(String[ ] args) {
        SpringApplication.run(MyApp.class, args);
    }
}
```

附录 C
注解式配置

在托管环境中，如程序以 war 包的形式部署于 Tomcat 服务器内，则可以采用如下注解对 Servlet、过滤器、监听器进行配置，免去了代码式配置。

- @WebSerlvet
- @WebFilter
- @WebListener

1. @WebServlet 举例

代码 附 C-1：@WebServlet 配置举例

```
@WebServlet(name="user.servlet", urlPatterns={"/user", "/users"})
public class UserServlet extends HttpServlet{
    public void doGet(HttpServletRequest req, HttpServletResponse res) {
        ...
    }
}
```

2. @WebFilter 举例

代码 附 C-2：@WebFilter 配置举例

```
@WebFilter("/user/*")
public class UserAuthenticateFilter implements Filter {
    public void doFilter(HttpServletRequest req, HttpServletResponse res) {
        ...
    }
}
```

3. @WebListener 举例

代码 附 C-3：@WebListener 配置举例

```
@WebListener
public class GlocalSetupListener implements ServletContextListener{
    public void contextInitialized(ServletContextEvent sce) {
        ServletContext sc = sce.getServletContext();
        sc.addServlet("book.Servlet", BookServlet.class);
    }
}
```

参考文献

[1] 沃斯. Spring 实战（第 6 版）[M]. 张卫滨，吴国浩，译. 北京：人民邮电出版社，2022.

[2] 张振华. Spring Data JPA 从入门到精通[M]. 北京：清华大学出版社，2018.